FACING INFINITY

FACING INFINITY

Black Holes and Our Place on Earth

JONAS ENANDER

Translated from the Swedish by Nichola Smalley

FOREWORD BY FRANK WILCZEK

THE EXPERIMENT

NEW YORK

Facing Infinity: *Black Holes and Our Place on Earth*
Copyright © 2024 by Jonas Enander
Translation copyright © 2025 Nichola Smalley
Foreword copyright © 2025 by Frank Wilczek
Page 344 is a continuation of this copyright page.

Originally published in Sweden as *Mörkret och människan* by Albert Bonniers
Forlag, Stockholm, in 2024. First published in English in the UK by Atlantic
Books, an imprint of Atlantic Books Ltd., in 2025. First published in North
America by The Experiment, LLC, in 2025.

All rights reserved. Except for brief passages quoted in newspaper, magazine,
radio, television, or online reviews, no portion of this book may be reproduced,
distributed, or transmitted in any form or by any means, electronic or mechanical,
including photocopying, recording, or information storage or retrieval system,
without the prior written permission of the publisher.

The Experiment, LLC | 220 East 23rd Street, Suite 600 | New York, NY 10010-4658
theexperimentpublishing.com

Every effort has been made to trace or contact all copyright holders. The
publishers will be pleased to make good any omissions or rectify any mistakes
brought to their attention at the earliest opportunity.

The cost of this translation was supported by a subsidy from the Swedish Arts
Council, gratefully acknowledged.

THE EXPERIMENT and its colophon are registered trademarks of The
Experiment, LLC. Many of the designations used by manufacturers and sellers to
distinguish their products are claimed as trademarks. Where those designations
appear in this book and The Experiment was aware of a trademark claim, the
designations have been capitalized.

The Experiment's books are available at special discounts when purchased in bulk
for premiums and sales promotions as well as for fundraising or educational use.
For details, contact us at info@theexperimentpublishing.com.

Library of Congress Cataloging-in-Publication Data available upon request

ISBN 979-8-89303-085-3
Ebook ISBN 979-8-89303-086-0

Jacket design by Beth Bugler
Text design and typesetting by Tetragon, London
Jacket photograph by Adobe Stock/reme80
Author photograph © Rickard L. Eriksson

Manufactured in the United States of America

First printing September 2025
10 9 8 7 6 5 4 3 2 1

The authorized representative in the EU for product safety and compliance is
Easy Access System Europe, Mustamäe tee 50, 10621 Tallinn, Estonia.
easproject.com | gpsr.requests@easproject.com

To Katinka

CONTENTS

Foreword by Frank Wilczek		*ix*
Prologue: The Ring of Light at the Edge of the Darkness		I

I AMID STARS, WAR AND DARKNESS

I	The Priest Who Wanted to Weigh the Stars	II
2	The Dark Heart of the Milky Way	25
3	The Astronomer by the 'Mountain of Death'	45
4	Einstein and the Blind Beetle	64
5	Beyond the Event Horizon	82

II BLACK HOLES IN THE DEPTHS OF SPACE

6	Dying Stars and Spacetime Vortices	101
7	A Cosmic Symphony	121
8	The Shadow Hunters	141
9	The Origin of the Giants	169

III BLACK HOLES AND OUR PLACE ON EARTH

10	Pōwehi and the Right to the Land	197
11	Black Holes and Climate Change	224
12	Black Holes Are Our Fathers	243
13	Hawking's Last Journey	262
14	Are We Living in a Black Hole?	290

Acknowledgments	315
Notes	319
Sources	335
Illustration Credits	344
Index	345
About the Author	354

FOREWORD
by Frank Wilczek

Black holes are the stuff of dreams, and of nightmares. By presenting us with weird and paradoxical predictions, sanctioned by compelling mathematical deduction, they challenge us to expand our minds and to exercise our imaginations. *Facing Infinity* is an especially attractive introduction to a wonderfully attractive subject.

We can use black holes metaphorically, and weave stories around them. In his science fiction classic *Gateway*, Frederik Pohl made brilliant use of that possibility. The novel records a series of psychotherapy sessions between an astronaut and his AI therapist. His problem? He is haunted by an image of his not-quite former lover. She lives on, frozen in time, at the surface of a black hole. As she crosses the event horizon, she looks towards him. Her frightened, strangely quizzical expression reflects her pained realization that his recklessness has doomed her. She will be living that moment for the rest of his life. That thought tortures him.

Some scientific perspectives might offer comfort. The lover's psychological perception of that moment, as opposed to the perception of a distant observer, is properly momentary. It does not occupy an inflated psychological time. Indeed, as you will learn in *Facing Infinity*, nothing extraordinary happens when one first enters a black hole (the unavoidable crunch comes later). Also, the distant observer's perception of the event will fade as its manifestation shifts from the visible into the infrared, microwave and radio bands of electromagnetic radiation.

X FACING INFINITY

Or maybe they don't. The persistence of memory and the parting of lovers are profound aspects of the human condition. Imaginative thinking can give us fresh perspectives on those abiding concerns, which is all we should ask.

Of course, the primary goal of black hole research is not to produce fresh metaphors. For scientists, the enduring fascination of black holes derives from their multifaceted grandeur.

In the introduction to his magisterial treatise *The Mathematical Theory of Black Holes*, the Nobel Prize-winning astrophysicist Subrahmanyan Chandrasekhar wrote:

> The black holes of nature are the most perfect macroscopic objects there are in the universe: the only elements in their construction are our concepts of space and time. And since the general theory of Relativity provides only a single unique family of solutions for their descriptions, they are the simplest objects as well.

This quote highlights a first aspect of black holes: their mathematical purity. Their remarkable physical properties were uncovered through a long struggle of thought, involving many years and many researchers. Albert Einstein discovered the fundamental equations in 1915, and a few weeks later Karl Schwarzschild found the matter-free solution that would come to describe the simplest (non-rotating) black holes. But it was only in 1939 that J. Robert Oppenheimer and Hartland Snyder understood the dynamical process through which black holes form. Much later, in 1963, Roy Kerr discovered the most general black hole solution, which allowed researchers to understand what happens when they rotate. Every step of the way, the mathematics revealed strange surprises. Perhaps the most surprising result of all is that, unlike ordinary stars, which come in many shapes and

FOREWORD

compositions, a black hole is completely defined by its mass and angular momentum. As John Wheeler put it, 'black holes have no hair'. As gravity imposes its will on the matter that forms the hole, all incongruities are smoothed and all memories fade away.

A second aspect of black holes, a comparatively recent achievement, is their observed reality. Circumstantial evidence began to emerge in the late 1960s, as astronomers tried to make models that would do justice to the astonishing properties of quasars and X-ray sources. But it was only in the twenty-first century that observers were able to get a good view of some black holes' immediate neighbourhoods, and even to produce images (through processing information about surrounding matter) of the distorted space around a few. Most amazing of all is the information flowing from observations of gravitational waves, starting in 2015. These waves often arise from processes where two black holes merge into one larger one. In comparing the structure of the observed signals with the intricate, highly detailed mathematical predictions based on general relativity, physicists not only justify the Einstein-Schwarzschild-Oppenheimer-Snyder-Kerr work but transcend it. For example, notably, before settling into the ideal Kerr solution, the objects resulting from a black hole merger will quiver, shake and 'ring' like a gravitational-wave bell.

Here again it seems fitting to quote Chandrasekhar:

> It is indeed an incredible fact that what the human mind, at its deepest and most profound, perceives as beautiful finds its realization in external nature.

A third aspect of black holes is their continuing ability to pose mysteries and inspire questions. The results mentioned above, including the uniqueness of black holes and the predicted structure of gravitational waves they generate, were derived within the classical theory of general relativity. But over the past one hundred years we've learned that the world is governed not by classical theory, but by quantum mechanics. Theoretical exploration of black hole quantum mechanics has been intense but inconclusive, while its observational counterpart has languished. One theoretical result that seems clear is that at the quantum level black holes support lots of fine structure – i.e., they are hairy after all. Or at least, they have lots of stubble. Might gravitational radiation, probed more deeply, reveal the quantum character of black holes? That is one of many tantalizing questions about black holes that continue to fuel research.

In this connection, it seems appropriate to quote Einstein:

> The most beautiful experience we can have is the mysterious.
> It is the fundamental emotion which stands at the cradle of
> true art and true science.

Facing Infinity freshly illuminates all three of these essential aspects of black holes: their mathematical purity, their observed reality, and their continuing ability to pose mysteries and inspire questions. Here you will find vivid narrations of what it would be like to get very near to or even dive into different kinds of black holes. Here too are Jonas Enander's personal encounters with both the charmingly various cast of people who are moving the frontiers of black hole knowledge and the fantastic instruments they use. Here you will also discover that there have been unexpected yet vitally important practical spin-offs from the curiosity-driven quest to better understand these objects and their surprisingly central role in our universe.

Having read all this, you will find your dreams enriched and your nightmares transformed.

FRANK WILCZEK

FOREWORD

· · · · · · · · · ·

FRANK WILCZEK won the Nobel Prize in Physics in 2004 for work he did as a graduate student. He was among the earliest MacArthur fellows, and has won many awards both for his scientific work and his writing. He is the author of *A Beautiful Question, The Lightness of Being, Fantastic Realities, Longing for the Harmonies, Fundamentals,* and hundreds of articles in leading scientific journals. His Wilczek's Universe column appears regularly in *The Wall Street Journal.* Wilczek is the Herman Feshbach Professor of Physics at the Massachusetts Institute of Technology, founding director of the T. D. Lee Institute and chief scientist at the Wilczek Quantum Center in Shanghai, China, and a distinguished professor at Arizona State University and Stockholm University.

Prologue

. . . .

THE RING OF LIGHT AT THE
EDGE OF THE DARKNESS

The light blinds you.

You raise your hands to shield your eyes, letting fragments of light filter between your fingers. White, blue and red stars are shining brightly in the darkness.·

You're floating in space. There's no up or down. There's nothing you can grip onto with your hands or push against with your feet. The only thing around you is a gaping expanse, a void. You are floating in it like a drop of water in the ocean.

Thick white gloves cover your hands. They are part of your spacesuit, your only protection from the deadly vacuum that surrounds you.

You lower your hands and squint. In the starfield in front of you, you see a darkness. There are no stars there. There's no light, nothing but cold, black emptiness: a black hole.

A black hole is a place in the universe with such strong gravity no light can escape it. That's why black holes are dark. But the darkness not only signifies the absence of light, it also represents a limit of knowledge. No particles, no radiation and no other information can exit a black hole. If you want to know what is going on in the darkness, you have to travel right into it.

You realize that the darkness in front of you has grown. You are falling towards it and there is nothing you can do to stop yourself. You have no spaceship, no rockets, no way to alter your course.

The darkness feels both menacing and alluring. Like the explorers of old, you are venturing into the unknown. You are going to find out what happens in one of the strangest places in the universe, a place no one else has ever visited.

But there's a difference between you and those adventurers: after they had explored far-away places, they could return home and describe what they had seen. You will not be able to travel back and tell your fellow humans what you've been through. Once you've fallen into the black hole, you can never turn back. Its gravitational pull is too strong. The darkness will swallow you forever.

'In space the universe grasps and swallows me like a point; in thought I grasp the universe.'[1] These are the words of the seventeenth-century French philosopher Blaise Pascal. He imagined that humans, with their capacity for thought, could gain comprehension and understanding of how the universe works. But the darkness of the black hole is so compact that the human mind can barely comprehend it. So how can you and other humans understand what a black hole is?

Perhaps an analogy will help.[2] Imagine you're floating in a river. It's night-time. The river is flowing towards a waterfall. With a flash of terror, you realize you must swim against the current to avoid being swept over the falls, but the closer you get, the faster the river flows. In the end, it's moving so fast you can't swim away. However hard you try, you are pulled inexorably towards and over the falls. The river is flowing too fast and you are swimming too slowly.

But let's assume there's someone else who can swim faster than you. At the point beyond which you're unable to escape the waterfall, this swimmer would be able to, though only to a

THE RING OF LIGHT AT THE EDGE OF THE DARKNESS 3

certain point. At a certain distance from the falls, this swimmer too would be dragged along by the water's flow, because the water would be moving faster than they can swim away. There is, therefore, a boundary in the river, set by the highest speed a human can swim. Beyond this boundary, no human can swim against the direction of the river's flow. The boundary itself is invisible; there's nothing in the water to indicate where this boundary is.

Now imagine that the river is so wide you can't see its banks. It's also so deep you can't see the bottom. Because you're swimming at night, the stars are all you can see. Substitute space for the river, and the speed of light for the maximum swimming speed, and you'll find yourself in a situation similar to your fall towards the black hole. Instead of being swept along by the water in the river, you're now being conveyed by the movement of space. If you'd had a spaceship, you could have turned around and travelled away from the darkness. But just as it's impossible to swim upstream after a certain point in the river, there is a boundary around the black hole after which you cannot turn back. It makes no difference how powerful the spaceship's rockets are. In the end, space is flowing so fast towards the darkness that not even light can travel in the opposite direction.

The boundary after which light cannot escape forms the surface of the black hole. It is known as the *event horizon*, and it is a surface spun from space and time.

It might sound odd that space can flow as a river does. But Albert Einstein realized more than a hundred years ago that this is possible. He discovered that matter and energy can distort space and time. Far from being static arenas in which our lives play out, space and time are active participants in the drama of the universe. But we'll come to that later. Right now all your focus is on the darkness around you.

You're falling into an abyss of space and time that has an enormous gravitational pull. But falling in space is not the same as falling on planet Earth. On Earth you can feel the air rushing past your face and hear your clothes fluttering in the wind. In the emptiness of space, however, there is no air and no sound. All you can feel is your spacesuit bumping against your body, and all you can hear is your breath, though you start to perceive a thudding noise that has grown louder as the darkness has deepened. You realize it's the sound of your heart, beating harder and harder the closer you get to the black hole. It's as though your heart fears what you will meet in the darkness, as though it knows you will have to sacrifice something in order to see what is happening inside.

The idea that knowledge comes at a cost features in many myths. When Eve took a bite of the fruit of the tree of knowledge, she and Adam were driven out of the Garden of Eden. When Faust made a pact with the Devil to gain unlimited knowledge of worldly things, he was made to pay with his soul. And when the Norse god Odin sought knowledge of the world and the future, he had to sacrifice an eye, throw himself on his spear and hang himself from the tree Yggdrasil.

Knowledge comes at a cost. The greater the knowledge, the higher the price. To find out what happens in one of the darkest and most peculiar places in the universe, you will have to pay the highest price of all: your life.

The point at which this will happen depends on the size of the black hole. The larger it is, the longer you can survive. The black hole you are falling towards right now is almost as big as the solar system. You can pass through its surface painlessly, but after that, your life will be over in only a few hours.

Your field of vision is increasingly taken up by the dark sphere. As the darkness grows, the light around it seems to change. You

THE RING OF LIGHT AT THE EDGE OF THE DARKNESS   5

see multiple copies of the light from the stars appear on either side of the black hole. At the same time, the stars' light seems to grow brighter, becoming compressed along the edge of the black hole. It occurs to you that the darkness is controlling the light. The black hole's powerful gravity is making the stars' light travel along peculiar paths that multiply its radiance. Phantom stars form in space, like mirages in the desert. These strange star-doubles disorient you. You want to tell your friends how the encroaching darkness fills you with terror, how the light of the stars is distorted and how helpless you feel. But even if you had a radio transmitter, you wouldn't be able to send your friends a message after you'd passed the event horizon. You will have to bear those final moments in the darkness alone.

Deep within the abyss, there is a point so extremely dense it is hard to wrap our minds around what goes on there. This point is called the *singularity*. In our river analogy, this is represented by the waterfall. As you were swept along by the river, you were unable to avoid the falls in the end. Similarly, everything that passes the event horizon will ultimately reach the singularity. There, all matter and all light will be concentrated in a state that is so distorted, even space and time seem to cease to exist.

You are gaining speed as you fall towards the darkness. There is nothing you can do to avoid it. You turn your head and look around you. The area of space behind you is growing darker and darker. You lose your ability to orient yourself. You cannot tell where you came from, or how far you are from the black hole. Are you already inside it? You don't know. There's no sign at the event horizon that says 'You are now passing the point of no return.'

The dark sphere seems to surround you in every direction. You flail your arms and legs in a desperate attempt to escape your journey towards the singularity, but it's pointless. There's no escaping the singularity. All that happens is that you start sweating.

You close your eyes, take a deep breath and think about what awaits you. When you travel feet-first towards a black hole, your lower body will feel a greater force than the upper body. You begin to be pulled apart. But it doesn't happen the same way as on a torture rack; instead, everything in your body is drawn out, from your skeleton, your tendons and your muscles, all the way down to your nerves, cells and DNA. Luckily, you'll feel almost nothing when it happens. From the moment you first notice the pain shoot through you, less than a second will pass before you are decimated. But unlike on Earth, where a person's dead body can be buried, there will be nothing left of you. Your body will be dissolved into the darkness.

The smell of your own sweat fills your spacesuit. You try to take deep breaths, but hear yourself breathing faster and faster. Pain flashes from your feet to your head. You tense every muscle in a final attempt to stop your body being strung out.

You open your eyes. The light of a billion stars blinds you. They are outside the black hole, but their light has been concentrated into a thin ring inside it. The ring is caught between the darkness in front of and behind you. You are in the centre of the ring.

Before you can even scream, your last second of life has passed. You've been torn apart by the darkness.

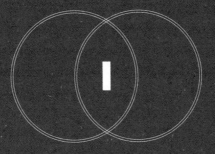

AMID STARS, WAR AND DARKNESS

The hub of the Milky Way is full of dust and stars, furious speeds and primeval patterns. The stars turn in an elliptical dance. Like fireflies in the night, they seem simultaneously to be on their own idiosyncratic journeys, while forming a fellowship in the darkness. One of the stars is whizzing round at immense speed. It is getting a taste for how it feels to approach the speed of light, to be slung around, deformed and then continue on its elliptical course. In just sixteen years it completes its orbit around the darkness that leads the stars' dance: an enormous black hole.

Chapter 1

· · · · · · · · · · · · · · ·

THE PRIEST WHO WANTED
TO WEIGH THE STARS

'm not sure any author wants to begin a book by taking their reader's life. But I wanted to explain what you would experience if you fell into a black hole. As luck would have it, neither you nor anyone else will experience the dramatic journey into an object like this – no human has so much as come close to one.

But even if no one has travelled towards a black hole, we are all moving *around* one. This cosmic giant is called Sagittarius A*, it has a mass equivalent to more than four million Suns and is located in the middle of the Milky Way. Just as the planets travel around the Sun, our solar system revolves around the centre of our galaxy. One revolution takes 230 million years, and carries us around billions of stars, vast clouds of dust – and the black hole Sagittarius A*. But don't worry, we're so far away there's no risk of us being pulled into it.

Today, astronomers know that there is a gigantic black hole like this at the centre of most galaxies. They also estimate that there may be more than a hundred million black holes in the Milky Way, created when certain kinds of stars died and imploded. A recent study even estimated that there could be as many as 40 quintillion (that is 40,000,000,000,000,000,000) black holes in the entire universe.[1] In this book I'll tell you how these black holes work, where they come from and what role they play in the universe. You will also discover how astronomers are able to confirm the existence of black holes, even though they are completely dark. And I will tell you about something that has truly astonished me: the influence they have on our lives here on Earth. By studying black holes in outer space, we can learn more about the conditions of our own existence. They can give us new insights into the relationship between light and darkness, creation and destruction, and even life and death.

The story of how humans began to explore the dark bodies of the cosmos opens on an unexpected scene: in a rectory in the village of Thornhill in the north of England, back in the late eighteenth century. A clergyman named John Michell lived here, and he became the first person to figure out that the gravity of a celestial body can be so strong that no light is able to leave it. He came to this realization when he set out to weigh the stars in the night sky, and it is with him that our tale about black holes begins.

THE LINNAEUS OF THE SKY

It's impossible to say whether John Michell was pleased when he started working as a rector (the formal title of a parish priest in the Church of England) in the Yorkshire parish of Thornhill in 1767. He left no diaries, and the few letters that remain tell us little about his emotional life. We hardly even know what he looked like. The only thing we have to go on is a letter from a contemporary of Michell's that describes him as 'a little short man, of a black complexion, and fat'.

What we do know, however, is that when Michell decided to become a priest, he gave up a successful scientific career. At the University of Cambridge he had conducted research into artificial

THE PRIEST WHO WANTED TO WEIGH THE STARS 13

magnets, the movement of earthquakes through the Earth's crust, and whether there was a structure to the apparently random distribution of stars across the firmament. He was a member of the prestigious Royal Society, read Ancient Greek and Hebrew, and was named Professor of Geology at the age of thirty-eight. He was, in short, a skilled scientist and polymath, respected and well liked by his colleagues.

Yet in spite of this, Michell elected to surrender his professorship after just two years. He did so for love. Michell wanted to get married, but according to university tradition, professors had to be celibate. We do not know what Michell thought of this requirement, but his actions speak loud and clear: he resigned and got married soon afterwards. He also sought the prestigious post of Astronomer Royal, which, in addition to a steady income, would have given him access to some of England's finest telescopes.

Michell was chosen as one of ten candidates for the appointment, but in the end he did not get the job. At around the same time, his wife died from the complications of childbirth. Suddenly, Michell was without both work and wife, and he had to raise and provide for his child alone. Seeking a more stable existence, he decided to give up his scientific career entirely and become a rector.

At the age of forty-three, Michell moved to Thornhill Rectory, to take up his new role. He appointed blacksmiths, plumbers and glaziers to restore the village's old stone church, and outside the rectory, he laid out a botanic garden, growing grapes, strawberries and exotic plants from such distant places as Mexico, India and China. He also got married again, to a local woman.[2]

Michell's life might seem idyllic, but around him major changes were in the works. The Seven Years' War, in which the great powers of Europe fought on five continents for control over their colonies, had come to an end, but conflicts continued to grow, within the great powers and between them. In France, poor harvests led to rampant criticism of the aristocracy and their land rights. On the other side of the Atlantic, the mood between the American settlers and the British Empire was tense. The French Revolution and the American War of Independence were on the way. Fundamental change was coming to the world Michell lived in.

But political and social anxieties were not the only harbingers of upheaval. Human experience of space and time was changing, and, with it, ideas of our place in the cosmos. The Scot James Hutton, now known as the 'father of geology', had studied the cliffs and sedimentary layers along the east coast of Scotland. He realized that they had been created by long-term geological processes, and that the age of the Earth must be counted not in the thousands, but in the millions of years (we now know that it is considerably older: 4.5 billion years). 'The mind seemed to grow giddy by looking so far into the abyss of time,' wrote one of Hutton's friends of these new geological time frames.[3]

As geologists were realizing that earthly timescales were vast, astronomers began to discover that the same applied to the distances of the cosmos. In the sixteenth century the Polish astronomer Copernicus had argued that it was not the Earth, but the Sun that was at the centre of the universe. In the eighteenth century, astronomers realized that this too was wrong. The Sun was but one star among many moving through that great assembly of stars we call the Milky Way. What's more, many Enlightenment thinkers also suspected that the centre of the cosmos was not even the Milky Way – they thought there were other galaxies in the depths of space. In the late seventeenth century, the French physicist and mathematician Pierre-Simon Laplace summarized this new cosmic understanding when he wrote, 'man now appears, upon on a small planet, almost imperceptible in the vast

THE PRIEST WHO WANTED TO WEIGH THE STARS   15

extent of the solar system, itself only an insensible point in the immensity of space'.[4]

It was as though humanity's place in the cosmos had shrunk. But Michell was not daunted by this. Throughout life, his scientific credo was ambitious. He wrote that he wanted to explore 'the infinite variety which we find in the works of the creation'.[5] He belonged to the Enlightenment era. It was a time characterized by the critique of dogmatic ideas and the enthusiastic exploration of the world through rational argument and experimental methods. The French Enlightenment philosopher and mathematician Jean le Rond d'Alembert wrote that the enthusiasm of this new era was 'like a river which has burst its dams'.[6]

As a follower of Enlightenment ideas, Michell participated in this wide-eyed exploration of the world, even after becoming a priest. Like a Linnaeus of the sky, he wanted to be a botanist of stars, mapping their distance, size and mass, and it was when he began to do so that he had a surprising realization: that there could be a limit to human knowledge, an impediment to the ambition of the Enlightenment. He realized that space might contain objects that were gigantic, dark and impossible to see.

DARK STARS

Michell made his discovery when he figured out a new method for measuring the distance to the stars. For centuries, astronomers had tried without success to determine how far away the closest stars were. They were applying a method known as *parallax measurement*, which can be illustrated with a simple example. Look at a nearby object, then hold out your arm and place your thumb in front of the object. If you close your left eye, your thumb will appear to be to the left of the object in front of you. If you now close your right eye, your thumb will be to the right of the object. The reason for this is that your eyes see your thumb from different perspectives, so the position of your thumb relative to the object appears to shift.

In a comparable way, a star's position in the sky can change as the Earth travels around the Sun. In my analogy, your left and right eye represent two observation points along the Earth's orbit, your thumb represents a nearby star you want to measure the distance to, and the distant object represents a star that is further away. As the Earth orbits the Sun, an astronomer observing a nearby star in the sky can see its position change relative to distant stars. With the aid of geometric analysis, it is then possible to figure out how far away the star is, if the astronomer's telescope is sharp enough. The further away a star is, the harder it is to observe the change in the star's position.

Though many had tried, no astronomer in Michell's time had successfully used parallax measurement to determine the distance to our nearest stars. This meant that they must be an extremely long way away. Since the parallax method had failed, Michell wanted to try to find a new way of determining the distance to the stars. Through a complex rationale, he realized that *if* it were possible to determine a star's mass, it would also be possible to determine how far away it was. But in order to do so, Michell needed to weigh the stars in the heavens. As he was unable to travel to the stars to study them, he set out to determine the stars' mass by studying the light they emitted. His starting point was the force that operates throughout space: gravity.

Gravity is a constant phenomenon in our lives. If we drop something, it falls towards the ground. If we walk up a hill, we feel tired. If we want to lift a heavy object, we undertake a tug-of-war between the strength of our own muscles and the strength of the Earth's gravity. But gravity is not just pulling the object. It

THE PRIEST WHO WANTED TO WEIGH THE STARS   **17**

also grounds our bodies on the Earth's surface. It ensures that the Moon orbits the Earth, and that the Earth and the other planets travel around the Sun. Gravity governs not only our lives, but also the fate of the whole solar system and, by extension, the whole universe. Therefore, if we understand how gravity works, we can understand how the entire universe developed.

In 1687 Isaac Newton published a book that profoundly contributed to our understanding of the characteristics of gravity. It was called *Philosophiæ Naturalis Principia Mathematica*, and in it he demonstrated that the gravitational force between two bodies – such as the Moon and the Earth, or the Earth and the Sun – is dependent on three things: the mass of one of the bodies, the mass of the other and the distance between them. The greater the mass, the greater the force, and the greater the distance, the smaller the force.[7] Thanks to this description of the relationships between these elements, Newton was able to explain a number of phenomena, such as the shape of the planets' orbits, how the Moon causes the Earth's tides, and why comets appear and disappear from the sky.

Michell realized that he could use Newton's results to weigh the stars, by studying their light. He reasoned that the larger the star's mass, the greater its gravity, and the greater a star's gravity, the more influence this force would have on the light it sends out. Michell made use of another of Newton's theories that stated that light was like small particles, or 'corpuscles', and he imagined that the light of the stars was affected by gravity like an object tossed in the air here on Earth. When we throw an object up, the force of gravity makes the object's speed decrease. In the end, it slows down so much it returns to the Earth's surface. Perhaps, Michell thought, the speed of light is affected in a similar way. He assumed that light was a particle travelling at immense speed, and imagined that when a star sent out these little particles, their speed would decrease depending on the strength of the star's gravity. Because gravitational force is dependent on a star's mass and size, it should be possible to deduce information about the star's characteristics by measuring how fast its light travelled.

To use modern terminology, Michell was talking about the concept of escape velocity, which can be explained by means of a simple experiment. Take an object you can hold in your hand. It could be a coin or a matchbox. Throw the object a few centimetres in the air. The object will travel upwards, before falling back into your hand, or to the ground. Now throw the object into the air again, moving your hand and arm faster this time. The object will reach a point higher in the air before it falls back down. After just two throws, you can surmise that the higher the speed at which you throw the object, the higher it will go before falling. This conclusion provides an important insight to help you understand Michell's thinking – and, ultimately, how black holes work.

If you were capable of throwing the object towards the sky with a velocity of eleven kilometres per second, it would never come back to Earth. It would travel through the air and eventually leave the Earth's atmosphere, continuing into space. The initial velocity an object would have to travel at to avoid falling back to the ground is called the escape velocity. Of course, it's impossible to throw an object so fast. Not even spaceships taking off from the Earth's surface need to reach that speed, because they accelerate using their rockets; furthermore, the escape velocity decreases the further they get from the Earth. But, despite this, the escape velocity is a good gauge of what it takes to overcome the Earth's gravity.

The gravitational force of objects in space varies, and therefore so does their escape velocity. From the Moon, the escape velocity is 'only' two kilometres per second. If you happened to be on an asteroid, the escape velocity could be as little as a few metres per second, which means you could leave the asteroid with a mere jump.

On more massive bodies the escape velocity can be much higher. At the surface of the Sun it is 615 kilometres per second. The stronger the gravitational force of a celestial body, the higher its escape velocity will be. On the basis of this realization, Michell came to a decisive conclusion: if a star has sufficient gravity, not even light will be able to leave it. 'All light emitted from such a body,' wrote Michell, 'would be made to return towards it, by its own proper gravity.'[8] Such a star would be completely dark.

Michell calculated how large such a dark star might be, taking the Sun as his starting point. What it consisted of and why it shone were not known at the time, though astronomers estimated that it was more than one hundred thousand times the size of Earth. Moreover, physicists had succeeded in measuring the speed of light, confirming that it was astonishingly fast, close to 300,000 kilometres per second. Michell calculated that a star with the same composition as the Sun, but with 500 times the diameter, would have such strong gravity that no light would be able to escape it. A star of this size would have an enormous mass – 125 million times that of the Sun, and its diameter would be larger than the orbit of Mars.

Michell's discovery indicated that there was more to gravity than objects falling to the ground, or planets orbiting the Sun. Gravity was suddenly placing a boundary on knowledge. These dark objects presented a challenge to science in its ambition to survey and understand the world.

Michell had no proof that these dark stars existed, however. They were merely an idea, a product of theoretical deduction and mathematical reasoning. But the intellectually restless Michell came up with a method of identifying these dark stars: observing other stars moving around them. Just as a lighthouse's beam testifies to the presence of otherwise-invisible cliffs on a dark night, the stars that move around these dark celestial bodies would reveal their existence.

What Michell felt when he thought about these gigantic, dark objects we will probably never know. But I imagine him walking out of his rectory one autumn night. He wanders through the long grass, passes a pond full of goldfish and makes for his telescope. Scents drift from his botanic garden on the cool night air. The sky is cloudless and the stars are shining brightly, and he leans in to his telescope, placing his eye against the eye-piece and turning it towards the star cluster Pleiades, in the constellation of Taurus. Michell has been studying the stars in the Pleiades, with their beautiful names – Pleione, Elektra, Maia, Alcyone – for many years. But tonight he turns his telescope away from the stars to gaze right into the darkness. Does he feel afraid at the thought of such huge, dark bodies existing in space? Or does he believe that we humans will one day be able to see other stars orbiting them?

A VERY CURIOUS PAPER

When I read Michell's letters and try to imagine his life, I'm struck by how multifaceted he was. He was a rector and a businessman, he planned and planted a botanic garden, he mapped out his region's coal stocks, he invented new navigational techniques for the British Navy, he mastered Ancient Greek and Hebrew, and he made contributions to geology, astronomy and physics. He seems to have possessed an unquenchable thirst for knowledge, but he also had a serious problem: he struggled to make his results known. He was isolated in Thornhill and missed his London friends. His research could be conducted in solitude, of course, but in Thornhill he hardly had anyone with whom to

THE PRIEST WHO WANTED TO WEIGH THE STARS   21

discuss his findings. He dearly wanted to travel to the capital to meet with other scientists, but was hindered by the expensive tolls on the roads. 'The expense of such a journey is more than I can afford every year,' he complained in a letter to a friend.[9]

To counteract his loneliness, Michell invited his friends to Thornhill. Many of England's foremost scientists travelled to the Yorkshire village and stayed at the rectory. Even the prolific scientist and inventor Benjamin Franklin, who would play a key role in both the American Revolution and the foundation of the new republic, took time out of a diplomatic trip to England to visit the intelligent rector.

But in spite of the visits, Michell hesitated to share his findings. His isolation in the countryside seems to have made him anxious that others would take credit for his work. When the physicist Henry Cavendish wrote to ask him to share his discoveries, Michell refused. Cavendish wrote, 'I am sorry however that you wish to have the principle kept secret.'[10] He kept on at Michell to publish his results, and Michell reluctantly conceded that he had given 'hints' of his findings when he met scientists in London, but that his indications had probably been 'too obscure to have the drift of them fully understood'. In short: he wished to keep his discoveries to himself.

Then at last, on 26 May 1783, Michell sent an article to Cavendish, asking for it to be read at one of the meetings of the Royal Society. In the accompanying letter, Michell wrote that 'it might perhaps be possible to find the distance, magnitude, and weight of some of the fixed stars, by means of the diminution of the velocity of their light'.[11] In order to reassure himself that he would receive proper credit for his results, he added of his method that 'as far as I know, [it] has not been suggested by any one else'.

Cavendish read the paper 'On the Means of Discovering the Distance, Magnitude, &c. of the Fixed Star' to the members of

22 FACING INFINITY

the Royal Society. It was the first official introduction of the idea that dark objects can exist in space without emitting any light. It took place as Mozart's *Great Mass in C Minor* had its premiere in Vienna, as the Montgolfier brothers undertook the first manned balloon trip in Paris, and as the Continental Army fought British soldiers in North America.

In Paris, Benjamin Franklin, who had visited Michell in Thornhill, was preparing to negotiate with Great Britain for US independence. Around the same time, Franklin received a letter from the President of the Royal Society, stating that Michell had written 'a very curious paper'.[12] Franklin was thus one of the first people outside Britain to hear about Michell's new ideas. We don't know whether Franklin discussed Michell's vision among Paris's scientific circles. But a decade later, Michell's ideas turned up once again in the French capital. Pierre-Simon Laplace, the leading physicist and mathematician in France at the time, had survived the terrors of the French Revolution with no more than a scare, though some of his colleagues were not so lucky: the chemist Antoine Lavoisier was beheaded on the guillotine. In 1796 Laplace published a multi-volume work in which he summarized all the astronomical knowledge of his time. He wrote that 'there exist then in space obscure bodies as considerable, and perhaps as numerous as the stars', and that 'it is therefore possible that the largest luminous bodies in the universe, may... be invisible'.[13] He called these bodies *corps obscurs* (dark bodies). Perhaps Laplace was inspired by Michell, since the President of the Royal Society had sent Michell's article to the Frenchman too.[14] Regardless of the source of Laplace's inspiration, it is fascinating that he laid out a whole cosmic vision in which these dark celestial bodies played a central role.

However, confirming the existence of these dark objects required observations. At the beginning of the nineteenth century

THE PRIEST WHO WANTED TO WEIGH THE STARS

Johann Georg von Soldner, a self-taught astronomer at the Berlin Observatory, read about Laplace's and Michell's deductions. He wrote that 'there may be heavenly bodies that, due to their size and the strong attraction associated with it, do not emit any light, or at least not to any distance; and that the largest bodies in our heavenly system must therefore remain invisible'.[15]

Von Soldner wondered whether there might be a gigantic, completely dark object in the centre of our galaxy. Just as the planets orbit the Sun, so too might the Sun be orbiting a central object in the middle of the Milky Way. Since no one had ever seen such a central object, reasoned von Soldner, it could conceivably be completely dark.

In twenty years, the idea of these dark celestial bodies had spread from Thornhill Rectory to the Royal Society's assembly rooms in London, then to the global metropolis of Paris, and finally to Berlin. These cities were key centres of scientific thought at the time. Three of the foremost physicians and astronomers of their day were discussing the existence of these objects and how they might be studied.

But just as quickly as the idea had arisen, it disappeared. 'It is very possible there may be no stars large enough to produce any sensible effect,' Michell wrote of the stars' ability to influence the light they emitted.[16] He doubted too whether gravity really affected light in the same way as other matter. He died in 1793 after a long illness, and was buried in the churchyard at Thornhill. Laplace crossed out the passage on these 'invisible bodies' when the time came to print a new edition of his volumes, and in Berlin, von Soldner dismissed the idea of a dark object at the centre of the Milky Way, concluding that no one had seen stars moving around such an object, 'even with the best telescope'.[17]

And so dark celestial bodies vanished from the scientific imagination. After all, they were merely a speculation based on

24 FACING INFINITY

certain assumptions: *if* gravity works as Newton had described, and *if* gravity affects light in the same way as particles, well, then a star of the same density as the Sun but 500 times the diameter would be completely dark.

But what if gravity didn't work that way out in space? Perhaps light didn't have the characteristics Michell and Laplace were assuming? At the beginning of the nineteenth century, light began to be viewed as a wave, rather than a particle, and it became unclear how such light waves might be affected by the gravity of the stars. Attempts to measure the speed of light coming from different stars hadn't produced any results either.[18] Michell's dark stars and Laplace's *corps obscurs* faded from the world of ideas.

In October 2020, more than two hundred years after the speculations of Michell, Laplace and von Soldner, a band of journalists are gathered at the Royal Swedish Academy of Sciences in Stockholm. The Secretary General of the Academy is about to divulge the recipient of this year's Nobel Prize in Physics. He begins by saying that the prize 'is about the darkest secrets of the universe'. The British mathematician Roger Penrose receives half the prize for his theoretical investigations into how black holes are created, while the German astrophysicist Reinhard Genzel and the US astrophysicist Andrea Ghez are awarded their share of the prize 'for the discovery of a supermassive compact object at the center of our galaxy'.[19]

The rector of Thornhill had been proved right. Gigantic, dark objects really do exist in space. Genzel and Ghez used Michell's method to identify the black hole in the centre of our galaxy: they observed how stars move around it. But what they have discovered is something much stranger than anything Michell could have imagined.

Chapter 2

· · · · · · · · · · · · · · ·

THE DARK HEART
OF THE MILKY WAY

I woke up at six o'clock to a message on my phone: *Lava is pouring out of the volcano!*'

Geoff Bower's voice booms out over the roar of the Jeep's engine. We're driving slowly up a potholed gravel track that leads to the top of the volcano Maunakea in Hawai'i. Bower has lived on the island for many years, and works as an astronomer at the Submillimeter Array, one of the many observatories at the top of the mountain.

Fortunately it wasn't Maunakea that was erupting. 'Although Maunakea is not technically an extinct volcano, it has not erupted in the last five thousand years,' Bower tells me. It was Kīlauea, on the south side of the island, that had had a minor eruption, and luckily no people or houses had been harmed.

Bower is driving me up Maunakea so that I can see the telescopes astronomers use to study black holes. As a project scientist for the Event Horizon Telescope Collaboration, he has been deeply involved in one of the most astonishing black hole observations ever made: the imaging of its darkness (more about that in Chapter 8). For my own part, I don't remember exactly when I first heard of black holes. It might have been some sci-fi show I watched as a teenager, or maybe a documentary on TV. But I've always viewed them as a kind of dark, threatening phenomenon that represents the unknown. At the same time, their darkness was also alluring.

When I started studying physics at university, my idea of what a black hole was expanded. They are objects that can actually be studied. With the help of mathematical formulae, we can understand their characteristics, and with sophisticated observations we can investigate what happens around them. This has led scientists to realize that they play an important role in the universe. They are not only destructive objects, they also have a creative function, governing the birth and development of galaxies, and, surprisingly, giving rise to the most intense light in space. They might even play an important part in the development of life in the universe. If scientists succeed in solving the mystery of black holes, we will be closer to understanding how structures are formed and how they dissolve, how space and time come into being, and how the laws of nature work. That's why I'm so fascinated by them, and it's why I'm sitting in Bower's Jeep heading up Maunakea.

I ask Bower why astronomers decided to build several of the world's most important telescopes in one of the most volcanically active sites in the world. 'First of all, the mountain is very tall,' says Bower, referring to the fact that its peak is more than 4,200 metres above sea level. At that altitude, light from space appears much clearer, as it does not need to travel as far through the turbulent layers of air that make up the atmosphere. 'Secondly,' Bower continues, 'Hawai'i is an isolated island in the Pacific Ocean. The wind blows smoothly over the telescopes, which reduces the twinkle of the light that passes through the atmosphere.'

I look out through the car window at the cloud cover below us. It's raining over the island, but above us the Sun is beating down. The higher we go, the more difficulty I have breathing. I feel dizzy and my lungs struggle to take in oxygen, and I feel no calmer

THE DARK HEART OF THE MILKY WAY

when I think of the sign at the bottom of the mountain that read: 'Altitude Sickness Can Be Fatal'. At the top of Maunakea, oxygen levels are significantly lower than they are at sea level. Although I'm finally about to see some of the world's greatest telescopes, all I really want to do is close my eyes and go to sleep.

We approach the peak at last. As I step out of the car, I think for a moment I've ended up on Mars. The gravel and stones on the ground are a reddish-brown colour. I can't see any plants or animals. Bower informs me that the comparison with Mars is not that far-fetched. On the neighbouring volcano Maunaloa, a billionaire has built an experimental facility for astronauts. It's where NASA tests the psychological effects of long-term isolation in a dry, stony, desolate environment reminiscent of our red neighbour. And NASA has road-tested lunar modules on Maunakea. Everything is barren and stark, everything except the thirteen observatories lined up along the peak, the white telescopes clearly silhouetted against the blue sky. My gaze sweeps across the ridge and I quickly identify the two telescopes the astronomer Andrea Ghez used to study the black hole at the centre of the Milky Way. I cast my mind back to the conversation I had with her a few weeks after she was awarded the 2020 Nobel Prize in Physics.

DOING THE IMPOSSIBLE

'When you travel to Maunakea and spend the first night at an altitude of four kilometres, your brain doesn't work particularly well,' says Andrea Ghez. 'It's a reminder of what a special place you're at.'

I'm speaking to Ghez remotely. It's evening where I am in Stockholm, but for her on the west coast of America the day has just begun.

Andrea Ghez is professor of physics and astronomy at the University of California, Los Angeles. Her interest in space was sparked early on, as she watched the TV broadcast of Neil Armstrong taking those first steps on the Moon in 1969. She dreamed of being the world's first female astronaut. Instead, she became an astronomer. She started studying physics, taking inspiration from women who had demonstrated that it was possible to follow one's passion, like Marie Curie (who was awarded the Nobel Prize in both Chemistry and Physics) or the pilot Amelia Earhart (the first woman to fly over the Atlantic). In the early nineties she completed her PhD at the California Institute of Technology.

'I don't think there's any stage of my career where I haven't heard the obnoxious thing "you're only getting to the next step because you're a girl or because you're a female",' Ghez tells me when I ask her what it's like to work in a field as male-dominated as astronomy. 'I was actually pretty used to hearing that, but I was also pretty used to thinking, "that's what you think, I'm fine!" I've been taught that if you have a good idea, just do it!'

One of the ideas Ghez had was to find out whether there is a black hole at the centre of our own galaxy. The centre of the Milky Way consists of many more stars than our region of the galaxy. If we lived on a planet that orbited one of these stars, the night sky would be full of stars that would have a combined brightness greater than that of the full Moon. The question was whether there was anything more than stars and clouds of dust in the galactic centre. In 1971 the astronomers Martin Rees and Donald Lynden-Bell had suggested that there might be a gigantic black hole lurking there.[1] Several astronomers studied the galactic centre with radio telescopes, and they detected a strong source of radio waves in the middle of this star-dense locality.[2] The radio signals were coming from a region that stretched approximately

THE DARK HEART OF THE MILKY WAY 29

a thousand times the distance between the Earth and the Sun. The radio astronomer Robert Brown called the source of this radio signal Sagittarius A* (or rather: its abbreviated form Sgr A*) because the galactic centre is in the direction of the constellation Sagittarius.[3] Brown borrowed the asterisk from the world of atomic physics, where atoms in a high-energy state are marked with that symbol. Brown thought the name was fitting because this region of the Milky Way emitted more energy than its surroundings. Personally I find it rather clumsy, because the name is pronounced 'Sagittarius A-star', despite the fact that it is not a star.

Sagittarius A* is right at the centre of the Milky Way.[4] It's important to point out that we are not only moving around the black hole, but around *all* the stars, dust clouds and other matter that are crowded into the dense centre of the galaxy. If Sagittarius A* suddenly vanished, it wouldn't affect our orbit.

The radiation from Sagittarius A* is generated by particles swirling about in the dust clouds of space.[5] Astronomers could see gas – mostly ordinary hydrogen – moving at high speeds around Sagittarius A*. From this they were able to estimate that there was a mass equivalent to several million Suns in the galactic centre. But exactly what this mass consisted of – whether it was stars, more gas, or perhaps just one or more black holes – they did not know.

Andrea Ghez wanted to find out which of these options it was. She set out to use the method John Michell had proposed more than two hundred years before: confirming the existence of a dark celestial body by studying how stars move around it. Just as ant tracks in a forest indicate the presence of a nearby anthill, the paths of the stars in space can indicate the existence of a dark, invisible object in space. But unlike the forest, where we can walk around in search of the anthill, we can't move freely in space. In order to prove the existence of the black hole, Ghez therefore had to observe the movement of the stars over many years.

There was one major problem. The Milky Way's dust clouds obscured the view. Radio signals are able to pass through these dust clouds, but seeing the stars deep in the centre of the galaxy was more difficult. When Ghez was finishing her PhD, two telescopes were being built on Maunakea that were capable of seeing through the cosmic dust clouds, right into the heart of the Milky Way. The first was completed in 1993, the other three years later. Together, they were named the W.M. Keck Observatory, after the oil magnate William Myron Keck, whose foundation had sponsored the expensive project.

The two Keck telescopes observe infrared light. Just like visible light, infrared light consists of electromagnetic waves. Visible light has a wavelength (the distance between the peak of one wave and the next) of approximately 400–700 nanometres (a thousand-millionth of a metre). It sounds small, but it's bigger than a virus or the molecules of our genetic code. The colours of natural light correspond to different wavelengths. Violet and blue have a shorter wavelength than green, yellow, orange and finally red, which has the longest wavelength of all visible light. Beyond red light, at wavelengths longer than 700 nanometres, is infrared light, which we can feel on our skin as heat. Just like radio waves, infrared light can travel through the dust clouds of space, and this enabled Ghez to see right into the centre of the Milky Way with the Keck telescope.

But when Ghez applied for permission to use the Keck telescopes, her application was denied.[6] No one knew whether it would be possible to see individual stars. Because the telescopes cost so much to use, it was safer to use them for projects with guaranteed scientific outcomes. 'It's a reflection of how conservative science is,' Ghez reflects, 'especially when resources are precious.'

THE DARK HEART OF THE MILKY WAY

But Ghez didn't give up: 'I find these moments where people say you can't do something energizing.' Like a politician on the campaign trail, she began to visit various institutes in search of support for her idea. Her campaign worked. In 1995 she and several of her colleagues flew to Hawai'i. They went up Maunakea to use the Keck telescopes. But their observation time was limited, while the low temperatures and lack of oxygen on the volcano's peak impaired memory and concentration. There was huge pressure to succeed. 'Every second is precious,' Ghez tells me. 'You might think of Hawai'i as a vacation place, but this is a time to be really serious. You're up at this extreme place, so every moment is focused on the work.'

Ghez wasn't alone in wanting to observe the movements of the stars. The German astronomer Reinhard Genzel, from the Max Planck Institute for Extraterrestrial Physics just outside Munich, had devoted his scientific career to studying the centre of the Milky Way.[7] He had one of the world's foremost telescopes to assist him: the Very Large Telescope, in Chile's Atacama desert.[8] A rivalry had emerged over which of the two research groups would be the first to observe the stars and identify what Sagittarius A* actually was.[9]

THE MULTIPLICITOUS PAST OF OUTER SPACE

When we look up at the night sky, we can see at most two thousand stars with the naked eye – and that's under the best conditions. It's a mere fraction of the 200 billion stars in our galaxy. In the previous chapter, we saw how John Michell and other astronomers strove to measure the distance to the stars. Today we know that these distances are vast. Our nearest star, which shines too faintly to be seen without a telescope, is Proxima

Centauri. It is more than forty trillion kilometres from our solar system. With numbers that large, it becomes awkward to use kilometres for stating the distance between stars and galaxies, so instead we use light years. In the vacuum of space, light travels at almost 300,000 kilometres per second. In a year, that makes 9.46 trillion kilometres, which is therefore the definition of a light year. It's an appropriate unit of distance, as it gives both the distance and the length of time the light has travelled before reaching us.

Proxima Centauri is 4.2 light years away from Earth. That means its light takes 4.2 years to travel here. By contrast, the light from the red star Betelgeuse, which shines clearly within the constellation Orion, has taken 640 years to reach us. Proxima Centauri and Betelgeuse highlight an important aspect of the night sky: when we see the stars, we are looking into the past of outer space. The light of some visible stars has travelled for thousands of years to reach Earth. Even when we look at the Sun, we are seeing it as it looked about eight minutes ago and the Moon as it was 1.3 seconds ago. So when we look up at the sky, we see a whole spectrum of the past. The multiplicitous *then* of space arches over our earthly *now*. Indeed, even on Earth we always see the past rather than the present. It takes time for light to travel from a source to our eyes. When we speak to someone standing in front of us, we are observing them as they were a few nanoseconds ago. We never see the world as it is, but as it has been.

Our solar system lies halfway between the centre of the Milky Way and its edge. From the galactic centre is takes a ray of light 27,000 years to reach Earth. Let's follow a ray emitted by one of these stars. When it began its journey, glaciers and ice shelves covered the northern and southern regions of our globe. Humans survived by hunting and gathering food, and might run into mammoths and sabre-toothed tigers on the snow-whipped plains.

THE DARK HEART OF THE MILKY WAY 33

Back to our ray of light, which is just departing the star-dense centre of the Milky Way. Its journey has no destination or aim, it's just doing what rays of light do: travelling straight through the vacuum of space. Along its many-thousand-year-long journey it passes other stars at a safe distance. These stars shine in all kinds of colours: some are blue, others red, some are white and some orange. Some of the stars are many hundreds of times larger than our Sun, others are significantly smaller.

When the ray has been travelling through space for 26,000 years, it reaches our region of the galaxy. Earth and humans have changed since it set out. Many of the glaciers, once kilometres high, have now vanished. Instead of hunting and gathering, humans have learned to grow and harvest. They have assembled into large groups and founded cities. When the ray has only 400 years of journeying left, humans make a major discovery. By placing a lens at either end of a tube, they can see further and more clearly: the telescope has been invented. In 1609 the Italian Galileo Galilei points this celestial eye-glass at the night sky and writes, rapt, that he can see 'great, unusual, and remarkable spectacles', such as the moons of Jupiter, the rings of Saturn, mountains on the Moon, and significantly more stars than could be seen with the naked eye.[10] In subsequent centuries, humans continue to improve the telescope. It gets bigger, its lenses are ever more precisely ground, its mirrors smoother. Instead of simply being observed by the human eye, the light collected by telescopes is saved on photographic plates and, later, with the aid of electrical circuits.

When the ray of light reaches our solar system, passes the orbits of the planets and travels through the Earth's atmosphere in a fraction of a second, Andrea Ghez is ready on the peak of the Maunakea volcano. With the aid of the Keck telescopes, she captures the ray to see what it can teach us about black holes.

'THEY BECOME LIKE YOUR LITTLE CHILDREN'

'There were questions about if you would see anything at all,' Ghez says. She and her colleagues took thousands of pictures of the galactic centre. Thanks to these many images, she was able to see individual stars, and the value of those intense nights on top of the volcano was clear. What remained unclear was whether the stars would help her prove there was a black hole in the centre of the galaxy; she needed to see whether they were moving. Only then could she confirm the existence of the black hole.

So Ghez travelled back to Maunakea to study the stars anew one year later. But there were technical issues and clouds covering the sky. Then, just as the constellation Sagittarius was about to disappear below the horizon, everything fell into place. The clouds dispersed for a moment, the equipment started playing ball, and Andrea Ghez and her colleagues were able to look right into the heart of the Milky Way. The greatest moment in her scientific career was upon her: the stars had moved.[11] 'It was incredibly exciting to realize how fast they were moving,' she recalls. 'Especially given that we had been turned down. It was a vindication.'

Ghez went on following the stars' dizzying dance for several years. Her images of the stars got even sharper, thanks to a technique known as adaptive optics that was installed in the Keck telescopes and had initially been developed by the US military for tracking enemy satellites. The technology used tiny mirrors inside the telescope that shift position a thousand times a second to correct light waves that have been distorted as they pass through the different layers of the atmosphere.

Ghez and her colleagues mapped more than a thousand stars.[12] 'They become like your little children,' she says, 'and it became crystal clear that So-2 was the star of the show.' So-2 is the

THE DARK HEART OF THE MILKY WAY 35

unremarkable name of a star that moves at a remarkable speed. As So-2 passed close to the conjectured black hole, it was whizzing at several thousand kilometres per second.[13] On Earth, we're able to fling individual particles about at such speeds using giant particle accelerators. In the galactic centre, something was doing the same thing with a whole star.

So-2 was making a circuit in only sixteen years, but it was impossible to see exactly what the star was orbiting. To find out what it was, Ghez needed something more than just observations. She needed a 400-year-old mathematical equation that had been formulated after the suspected murder of one of the foremost scientists in Europe.

THE WAR ON MARS

In 1599 the fifty-two-year-old aristocrat Tycho Brahe arrived in Prague.[14] He was the leading astronomer of his time, but he'd been driven from his homeland of Denmark by political intrigues. He was about to take up a new position as astronomer to the court of Emperor Rudolf II. For Brahe it was a prestigious post at the absolute centre of the powerful Holy Roman Empire, a polity that stretched all the way from the Baltic Sea to the Mediterranean.

In his luggage, Brahe had books of astronomic tables containing data from decades of precise observations of the heavens. Brahe had made these observations on the island of Ven in the Øresund strait, where he'd built the magnificent castle Uraniborg and the observatory Stjerneborg. With the aid of new instruments (including a quadrant and an armillary sphere, though not a telescope as this had yet to be invented), Brahe could map the shifting positions of the stars, the planets, the comets, the Moon and the Sun.

Kings and nobles from across Europe had travelled to Ven to see Brahe's magnificent observatory, but by the time Brahe arrived in Prague, Uraniborg and Stjerneborg were in ruins. The new Danish king, Kristian IV, and his closest advisors had driven Brahe out of Denmark and razed his castle and observatory. 'My fatherland now lies wherever, humbly, men behold the stars,' Brahe wrote soon afterwards, in what sounds like an attempt to cheer himself up.

Brahe wanted to continue his observations in Prague, but he also wanted to use his astronomic tables to solve one of the biggest mysteries of his time: what do the planets turn around? Is the Earth the centre point for the motion of the planets and the Sun too, or do all the planets, including the Earth, orbit the Sun?

Brahe believed it was possible to determine the structure of the solar system, but doing so required precise, methodical observations made over a long period of time – the kind of observations he had devoted his life to. He was primarily interested in finding out whether the model he himself had developed was correct. It stated that the Earth was at the centre of the universe, while the other planets orbited the Sun. In turn, the Sun and these other planets orbited the Earth. Brahe's model was a kind of compromise between the geocentric model, which placed Earth in the middle, and the heliocentric model, with the Sun at the centre. This latter model had been posited by the Polish astronomer Copernicus, among others, several years before Brahe was born.

But Brahe had a problem. Confirming that his model was correct required him to undertake complex calculations. These were so comprehensive that he didn't want to do them himself. Instead, he appointed a mathematically talented, but contrary, astronomer from Germany. His name was Johannes Kepler, and Brahe would soon regret that he'd ever had anything to do with him.

THE DARK HEART OF THE MILKY WAY

Kepler was born in 1571 in Weil der Stadt in southern Germany. He had a difficult upbringing. His father was absent fighting as a mercenary in the Netherlands, and at times he was cared for by his maternal grandparents, to Kepler's horror. When he caught smallpox as a child, his grandmother bound his hands together to stop him scratching the weeping sores. The sores spread around his eyes, leading to permanent damage that meant Kepler saw multiple images of the same object. In contrast to Brahe, whose vision was impeccable, Kepler was also shortsighted. 'For observations indeed I am by sight stupid,' Kepler complained.

In many ways, Brahe and Kepler were the exact opposite of each other. Brahe was a rich aristocrat; Kepler came from a poor family. Brahe was powerfully built and loved alcohol, he fenced (he wore a prosthetic nose made of copper since his own nose had been cut off in a duel), and had a wide social circle. Kepler kept himself to himself, was slight of build, and often suffered from fevers, headaches, inflammations, skin rashes and stomach cramps. He shunned both alcohol and dinner invitations.

They did, however, have one thing in common: just like Brahe, Kepler had been forced to flee his homeland. When religious discontent boiled over in Germany, the Lutheran Kepler refused to convert to Catholicism. Their lives in peril, he and his family were forced to flee. He ended up in Prague, charged with helping Brahe answer the question of how the solar system was structured.

Kepler soon grew dissatisfied with his new job. He too had created a model of how the solar system might look. Deeply influenced by mysticism and astrological thought, he was driven by a desire to find order behind the myriad forms that appear in nature. He imagined that the Sun was in the middle, and that a series of geometric relationships between tetrahedra, cubes and other so-called platonic bodies determined the planets' distance to the Sun. These relationships, he believed, reflected a divine mathematics. 'Geometry is one and eternal shining in the mind of God,' wrote Kepler.

Just like Brahe, Kepler wanted to know how the planets moved through space, and whether it was the Earth or the Sun that formed the midpoint of the universe. But unlike Brahe, Kepler didn't have access to the astronomic data needed to investigate which model was right. Brahe only shared the data needed for the calculations relating to his own system, and Kepler grew increasingly frustrated. To test his own models, he particularly needed Brahe's data about Mars, as the planet's movements were difficult to reconcile with the perfect circle Kepler imagined it ought to describe. Because Mars posed such problems for Kepler's model, he called his study 'The War On Mars'.

Kepler moaned about the boring calculations, and nagged Brahe to share his data with him. But Brahe refused. He had suffered plagiarism previously, and had no wish to repeat that bitter experience. 'He abounds in riches,' Kepler exclaimed, 'which like most rich people he does not rightly use.' The conflict between them grew, and in the end Brahe tired of it. 'I have decided therefore to have nothing afterward of commerce with him,' he wrote of Kepler, adding, 'whether through letters or orally, and I might wish that I never had any.'

In the end, though, Kepler got his hands on Brahe's tables – not because Brahe acquiesced, but because he was dead. Tycho Brahe died on 24 October 1601. To this day it's unclear what he died of, and soon after his death, speculations arose that it could have been murder. The Norwegian bishop of Bergen wrote of Brahe's death that 'an unpleasant rumor has developed, namely that he died, but not a usual death'.[15] Kepler, however, said that Brahe had died of natural causes after a dinner involving too much alcohol. Out of respect for his company, Brahe declined to leave the table to use the toilet, and this, according to Kepler,

THE DARK HEART OF THE MILKY WAY

caused him to develop a urinary infection that led to his eventual death. But in 1993 studies were carried out on a strand of hair from Brahe's grave. The tests showed high levels of mercury – had he been poisoned? Suspicion turned on Kepler.[16] He had a clear motive: to get Brahe's astronomical data. However, a subsequent investigation showed that the mercury content was nowhere near as high as initially claimed. The allegations against Kepler were dismissed, but the very fact that they were made shows how deep the conflict between Kepler and Brahe was – all stemming from the question of how space was structured and who would get access to the best astronomical data of their time.

One thing is clear: Kepler profited from Brahe's death. A few days after the Danish astronomer's demise, Kepler took over Brahe's post as mathematician to the court. He also kept Brahe's tables, in defiance of the attempts of Brahe's family to control them as part of the inheritance. Kepler took the data, putting all the materials in his bag and leaving Prague. 'I had possession of the observations and I refused to hand them over,' he wrote to a friend. Kepler would go on to use the tables to make one of the greatest discoveries in the history of astronomy, a discovery that would later enable astronomers to identify black holes.

For several years, Kepler struggled to make sense of what Brahe's data said about the structure of space. He sought mathematical and geometric connections. His approach might seem peculiar today – imagining that the Earth had a soul and that the movement of the planets in space resounded with a cosmic music – but it led him to a decisive discovery: that the orbits of the planets are governed by a fundamental mathematical structure.

Kepler presented his results in two works: *Astronomia Nova* in 1609, and *Harmonices Mundi*, or *Harmonies of the World*, in 1619. He showed that Copernicus was right about the planets orbiting the Sun, but that these orbits are not circular. Instead, they move

40   FACING INFINITY

along ellipses. He also showed that the Sun is not at the centre of these ellipses, but in one of their two focal points. But Kepler's discoveries didn't end there. The solar system has six planets that can be seen with the naked eye: Mercury, Venus, Earth, Mars, Jupiter and Saturn. These planets take different lengths of time to complete their orbits around the Sun. Kepler discovered that there was a deep mathematical connection between a planet's distance from the Sun and the time it takes for it to complete an orbit.

Let's take inspiration from Kepler and examine this relationship, but instead of Tycho Brahe's astronomic tables, we will use modern data on the planets.[17] It takes Saturn 29.4 years to orbit the Sun. Jupiter takes 11.9, Mars 1.9, Earth 1 year and Venus 225 days. Mercury, which is closest to the Sun, completes its orbit in a mere 88 days. A clear pattern emerges: the closer to the Sun a planet is, the shorter its orbital period.

Kepler discovered that, far from being random, these periods depended on how far the planets were from the Sun. We can use the Earth's distance from the Sun as our unit here. This measurement, which is called an astronomical unit, therefore has a value of 1 for Earth. If we measure the orbital period in years, our value will once again be 1. No matter how we multiply or divide these numbers by one another, the result will always be 1. But for Mars – the planet that had given Kepler such headaches – it was different. Mars is around 1.5 times as far from the Sun as Earth is. What Kepler discovered was that if we multiply this distance by itself three times, and then divide this figure by Mars's orbital period multiplied by itself twice, we get a result very close to 1: 1.0006. If we do the same thing for Venus, we get 0.99983, and for Jupiter, 1.0015. We can carry out this mathematical operation for all the planets in the solar system. We will always get a value close to 1, regardless of the planets' different orbital periods and

THE DARK HEART OF THE MILKY WAY   **41**

distances from the Sun. A deep mathematical structure underlies the solar system!

'First I believed I was dreaming,' wrote Kepler of his reaction to discovering this mathematical structure. The data he had access to was not as good as what we have, yet he managed to work out this relationship. Today this is known as Kepler's Third Law (the First Law concerned the planets' elliptical orbits, as mentioned previously, and the Second Law, which we won't explore here, concerns those planets' speed along these orbits).

When, as a student of physics, I read about Kepler's laws, they affected me deeply. The fact that a relatively simple formula could describe the planets' orbits shows how Kepler had achieved one of the most important goals in physics: uncovering the mathematical structure that characterizes the multiplicity of phenomena we observe in nature. I remember how fascinated I was to find out that it was the planets' movements that allow us to understand this structure. I found it even more exciting that, even though Kepler's thinking was inflected with mysticism and astrology, modern physics still bears the stamp of his search for fundamental symmetries and geometric relationships. It was as though humanity's search for meaning had hit upon a rich source, in spite of an approach that – to many modern eyes – seems strange.

But where does this value of 1 that the planets share come from? There's actually nothing particularly special about this value. In our calculation we used the Earth's distance from the Sun as a unit of distance and expressed the orbital periods in years. We could just as well have measured the distance in kilometres and the orbital periods in seconds. That would have given us a different value, but it still would have been common to all the planets.

Still, there must be something underlying this common value. Kepler never found out what it was. He died in 1630, poor and relatively unknown. Even in death, the contrast between the rich empiricist Brahe and the poor mystic Kepler was clear. Brahe's lavishly ornamented grave can still be seen in the Church of Our Lady Before Týn in Prague. As for Kepler, neither his body nor his grave remain. Two years after his death, his grave was destroyed as the Swedish soldiers of Gustavus Adolphus passed the city of Regensburg and the churchyard where Kepler was buried. His earthly remains disappeared, but his insights into the structure of space survived.

It was Newton who finally managed to explain what caused this mathematical relationship between the planets' distance from the Sun and their orbital periods. He demonstrated that this pattern was governed by the Sun's gravity, which is determined by its mass. This mass could therefore be determined by measuring a planet's orbital period and its distance from the Sun, and astronomers were able to calculate that the Sun's mass is approximately 2×10^{30} kilograms. That's an enormous figure: a two followed by thirty zeros. It is equivalent to the combined mass of more than 333,000 Earths.

Thanks to Kepler and Newton, it became possible for astronomers to calculate the mass of distant objects in space. All they need to do is observe one object's motion around another. This could be a moon orbiting a planet, planets orbiting other stars, stars orbiting each other, or, for Andrea Ghez and Reinhard Genzel, stars moving around a black hole.[18]

To ensure a scientific result is as robust as it can be, two or more independent teams of researchers should arrive at the same conclusion. That's how it was with Sagittarius A*. Separately, Ghez and Genzel and their colleagues used the stars' elliptical motion around Sagittarius A*, combined with Kepler's Third Law, to confirm both that its mass was equivalent to around four million Suns, and that this mass was concentrated in a very small

THE DARK HEART OF THE MILKY WAY

volume in the middle of the Milky Way. Over the years, Ghez and Genzel met at conferences, critiqued and discussed each other's results, and came, ultimately, to the same conclusion: Sagittarius A* is probably a black hole.

OUR COSMIC SURROUNDINGS

The Sun begins to set over Maunakea. Bower has guided me around the mountaintop and explained what each of the telescopes do. Unfortunately, we're not able to visit the Keck telescopes themselves. 'There are fewer astronomers here after the Corona pandemic,' Bower tells me. 'We learned how to make all observations at a distance. Only a small number of people are needed at the telescopes, and that makes it harder to visit them.'

But I'm happy all the same. On top of the mountain, it's possible to see how the universe becomes comprehensible, how advanced telescopes allow astronomers to follow the movements of the stars and survey our galactic home.

Bower drives me back down the mountain, dropping me off at the Maunakea Visitor Centre, at 2,800 metres altitude. Then he makes for home while I stay on to see the night sky. After all, Maunakea is one of the world's best places for stargazing. As night begins to fall, more and more people make their way up to the visitor centre. They share my desire to look up at the sky. I look at the Southern Cross, with its four prominent stars that have guided centuries of seafarers. It's the first time I've seen this famous constellation.

I see a cloud begin to build at the horizon. After a while I realize it's not a cloud, but the diffuse light of the Milky Way's billions of stars. I stare at it for a long time. Seeing the Milky Way this clearly gives me a sense of the vast scale of our cosmic surroundings. It's a turbulent feeling I struggle to put into words. Somewhere deep within that muddle of stars is Sagittarius A*, the galactic midpoint we are moving around right now. I try to picture in my mind how our solar system moves around it, and to reconcile this with the fact that the last time we were in this part of the galaxy, dinosaurs were roaming the Earth.

By tracking the stars' movements around Sagittarius A*, Andrea Ghez and Reinhard Genzel succeeded in doing what the rector John Michell had dreamed of more than two hundred years previously.[19] But there is an important aspect of their discovery that we have yet to touch on fully. How could they know that Sagittarius A* was really a black hole? To answer that question, we must take a step back and go into a little more detail about what a black hole actually is. So we'll leave Maunakea behind and make our way to one of the hardest-won battles of the First World War. It was there that the mathematical formula that describes a black hole was first discovered.

Chapter 3

. . . .

THE ASTRONOMER BY THE 'MOUNTAIN OF DEATH'

t's cool and quiet inside the archives of the University of Göttingen. The spring light floods through a large window and lands on the desk in front of me. On the desk is my laptop, a magnifying glass and several green folders.

I'm tired and my body is stiff. I slept poorly on an uncomfortable sofa bed in a spartan room I rented from two students. But my bad night's sleep wasn't only due to the sofa bed; I'm nervous too. I travelled two days by train from Stockholm to Göttingen, in Germany, driven by a question: how did humanity's search for black holes begin? In Chapter 2, we saw how, in the later eighteenth century, John Michell and Pierre-Simon Laplace introduced the idea that there might be stars with such strong gravitational force that they emit no light. But black holes are much stranger than that. They are not stars with powerful gravity, but space and time that have been altered so much that there are no ways out of them. I want to know what circumstances would even enable someone to imagine that such a thing was possible, and I'm hoping to find the answer in the folders in front of me.[1]

I open the first folder. It contains yellowed letters, postcards, notes – and the German astronomer Karl Schwarzschild's military ID card. It was filled out in the city of Namur in Belgium on 18 March 1915, and states that Schwarzschild was born on 9 October

1873 in Frankfurt am Main, that he is forty-one years old and 168 centimetres tall. His occupation is given as 'Astronomer'. It was while stationed in close proximity to some of the most intensive fighting in the First World War that Karl Schwarzschild discovered the mathematical formula for black holes.[2]

On the ID card is a photo showing him sitting in front of a staircase, dressed in a military overcoat and hat. A bushy walrus moustache sprouts from beneath his nose. He looks straight into the camera, eyes shining with a kind of inquisitive intelligence, as though to let you know he can solve all the puzzles of space with the calculations on the piece of paper clutched in his hand. During the war it was paper and pen, rather than a rifle or a bayonet, that were to become Schwarzschild's weapons.

Military identification card of the German astronomer Karl Schwarzschild. He served at a military weather station and as a lieutenant in an artillery unit during the First World War. In the late autumn and winter of 1915 he was stationed in Mulhouse near the German–French border. While there, he discovered what was later understood to be the formula for black holes (see image on page 49).

When the war broke out, Schwarzschild was the director of the Astrophysical Observatory Potsdam, just outside Berlin, and one of Germany's foremost astronomers. Although he was already forty, he decided to leave the comfortable life and volunteer to serve in the German army. The reason was his Jewish ancestry: Schwarzschild had been the victim of anti-Semitism, and he hoped that if he showed a willingness to fight for his nation, he and his kin would enjoy greater respect.

In the army, his talents were in great demand. The skills of an astronomer, such as minute optical observations, predicting the course of planets and comets, and knowledge of the atmospheric conditions on the Earth and the Sun, turned out to have significant military value. Schwarzschild worked at a military weather station in Belgium, before being relocated to an artillery unit. He was stationed on the Eastern Front in Lithuania for a month towards the end of the summer of 1915, before continuing to the town of Mulhouse on the German–French border, on the Western Front. At that time he was a lieutenant in the 10th Imperial Prussian Foot Artillery.

'It is unpleasant to constantly be reminded of the slaughter by huge explosions as I'm sitting here writing or working on my calculations,' Schwarzschild wrote to his wife Else a few days before New Year's Eve 1915. On Hartmannswillerkopf, in the Vosges region near Mulhouse, bitter struggles were raging between French and German soldiers. During the war the mountain became a killing field where tens of thousands of soldiers fell to bullets, grenades and shell fire, hunger and the cold, illness and lack of water. The Germans called Hartmannswillerkopf the 'Mountain of Death', the French called it the 'Man-Eating Mountain'.[3]

But Schwarzschild wasn't on the battlefield; his job was to calculate the effect of humidity, rain and wind on German artillery.

In a classified report, he wrote that the aim of these calculations was to 'reduce [unnecessary] shooting, save ammunition, and improve the element of surprise over the enemy'.[4]

Schwarzschild didn't restrict himself to military calculations. 'Today, as a Sunday treat, I have been studying Einstein's gravity theory,' he wrote in a letter to his wife in December 1915, adding, 'perhaps he, like Kepler, is closer to our Lord God, who reveals to him how He has made everything.'[5]

A few weeks previously, Einstein had presented the results of many years' work to the Royal Prussian Academy of Sciences in Berlin. He had been seeking to understand what gravity was. It may sound trivial; after all, everyone knows that gravity is about objects falling to the ground, or the feeling of a force pulling us towards the Earth's surface. But Einstein had explained to his colleagues at the Academy of Sciences that something much deeper was hidden behind our everyday experiences. Gravity, Einstein asserted, was actually about the fundamental properties of time and space.

Schwarzschild started studying Einstein's new theory, the one known as the general theory of relativity, in detail. 'It really is very peculiar,' he wrote to Else, noting excitedly that, 'if correct, it will be as important as this entire war.' As Schwarzschild was attempting to puzzle out what general relativity could mean for the characteristics of stars and the planets' motion around them, he discovered the mathematical formula that acted as the catalyst for humanity's search for a black hole, a search that has taken us to the very edges of space and time.

Schwarzschild discovered this formula as he listened to the thudding of cannons and saw the smoke over Hartmanns-willerkopf. While German troops were leading an attack that would take the lives of almost two thousand French soldiers, Schwarzschild was posting a letter to Einstein at Haberlandstrasse

THE ASTRONOMER BY THE 'MOUNTAIN OF DEATH' **49**

5 in Berlin. Schwarzschild wrote, 'As you see, the war is kindly disposed towards me, allowing me, despite fierce gunfire at a decidedly terrestrial distance, to take this walk into your land of ideas.'[6]

In this and the next two chapters, we will follow Schwarzschild on his wanderings through the world of Einstein's ideas. We will see what new insights Einstein had as to the characteristics of time and space, how this led to Schwarzschild's formula for black holes, and what it actually means.

Our tale begins with the opposite of darkness: light. It was while analysing light that Einstein first began to explore the characteristics of space and time, which in turn led him to make one of the greatest discoveries ever made by humankind. It was a discovery that concerned the absolute fundamentals of our, and indeed the whole universe's, existence.

Excerpt from a letter sent by Karl Schwarzschild to Albert Einstein on 22 December 1915. The letter marks the starting point for our understanding of black holes. The equation that Schwarzschild found describes how space and time change around a massive object, and is today known as the Schwarzschild metric.

AN ENCHANTED GLOW

Light plays a crucial role in our lives. We need light to see, life on Earth has flourished thanks to the light of the Sun, and with light from stars, we can study the properties of space. But what actually is light? When Einstein explored this question, he hit upon an answer that made him not only one of the greatest scientists in history, but also one of the most famous. Today, our most common image of Einstein is the one of the cheeky old man with a shock of white hair, sticking out his tongue at the camera and cracking witty pearls of wisdom like, 'I never think of the future – it comes soon enough'. But when Einstein was coming up with his most significant brainwaves, he was a little-known civil servant at the patent office in Bern, Switzerland, who had failed to achieve his dream of an academic career. He was unkempt, rebellious, and had an extraordinary talent for focusing on – and solving – theoretical problems. One of these problems was finding out how fast light travels.

The central role of light in Einstein's life was due, in part, to his upbringing.[7] Einstein was born in the small town of Ulm in southern Germany on 14 March 1879 – the same year in which Edison invented the lightbulb he is now famous for, the first female students were admitted to Oxford University, and Ibsen's *A Doll's House* premiered in Norway. Einstein's father Hermann came from a family of Jewish tradesmen, while his mother Pauline's family were grain merchants. They wanted to call their son Abraham, after his grandfather, but were worried that the Jewish-sounding name would cause problems due to society's prevailing anti-Semitism. In the same year Einstein was born, for instance, the organization Antisemitenliga was founded, with the express aim of countering Jewish influence on German culture.[8] So as not to jeopardize their son's prospects,

THE ASTRONOMER BY THE 'MOUNTAIN OF DEATH'

Einstein's parents therefore decided to name him Albert. They could hardly have guessed that their newborn son would go on to change humanity's understanding of the world.

A year or so after Einstein was born, the family moved to Munich, where his father set up an electronics company with his brother Jakob. The company, J. Einstein & Cie, began producing electric lights. The new technology had been illuminating ever-larger parts of society since the mid-nineteenth century. Arc lamps, gas discharge lamps and eventually Edison's cheap, practical lightbulb, too, shone on the streets, and in salons, factories and homes. All over Swabia in southern Germany, Hermann and Jakob Einstein had been installing electric lights in breweries, hospitals and inns. They became so successful they were rewarded with the commission to provide the electric lights for the 1885 Oktoberfest in Munich. For the first time, the Bavarian beer drinkers drank by electric light instead of gas or oil lamps. A local paper described how the Einstein family's new illuminations cast 'an enchanted glow over the festival site with its thousands of visitors'.[9] Perhaps six-year-old Albert helped his father install the lights.

As electric lights were beginning to illuminate the streets and squares, scientists were working hard to fathom what light really is. In 1864 the Scottish physicist James Clark Maxwell formulated a set of equations describing how electricity and magnetism worked. Using these equations, Maxwell could show that light consists of electrical and magnetic waves. There was, however, something odd about Maxwell's equations. They did indeed state that light should move at exactly the speed that had already been observed (almost 300,000 kilometres per second), but they also predicted that this speed should be the same irrespective of how someone who measures the speed is moving. This clashed with the standard view of many physicists, who were used to viewing speed as a relative phenomenon. For example, if you are standing still, your speed is zero relative to the Earth's surface. But all the while, the Earth is moving around the Sun at a speed of thirty kilometres per second. The solar system is in turn travelling around the Milky Way at a speed of 240 kilometres per second (when my editor read this, she exclaimed, 'I've been sitting here on my sofa for a while now, trying to get my head around the fact that I've just travelled 240 kilometres... and another 240... and another!').

So what is your speed as you're standing there? Zero kilometres per second? Thirty? Two hundred and forty? The answer is all of them. Relative to the ground, you are indeed at rest, but relative to the centre of the solar system, you're moving at thirty kilometres per second, while at the same time travelling at the dizzying speed of 240 kilometres per second in relationship to the Milky Way centre. By the time you reach the end of this sentence, it means that you'll have travelled almost a thousand kilometres through space, even though it feels like you haven't moved at all.

Speed is about how two things move in relation to one another. But Maxwell's equations stated that it doesn't matter if the one who measures the speed of light is at rest or moving fast relative to a ray of light. The result will always be the same: light moves at almost 300,000 kilometres per second *for everyone*. That seemed unreasonable. Imagine you're sitting on a train – you're completely still in relation to the train, but for someone standing on a platform watching the train pass, both you and the train have a high speed. If you, still on the train, switch on a torch and point it in the direction of travel, a ray of light will move away from you at almost 300,000 kilometres per second. But for the person on the platform, the light should, by rights, travel away from them at an even higher speed, because the light's speed relative to them ought to be the sum of the train's

THE ASTRONOMER BY THE 'MOUNTAIN OF DEATH'   **53**

speed and the light's speed as measured by you. After all, speed is a relative quantity! But Maxwell's equations seemed to say that light has the same speed both for you on the train and the person on the platform. But if you and the person on the platform have different perceptions of the train's speed in relation to yourselves, how could you have the same perception of the speed of the light on the train? How can the speed of light be absolute and all other speeds relative?

Even as a teenager, Einstein was obsessed with trying to understand this, and perhaps his upbringing led to him finding the solution. He had left the Luitpold Gymnasium in Munich, in part because he couldn't stand the Prussian discipline. He was also worried he would be forced to sign up for German military service, so he gave up his German citizenship. At the age of sixteen he became stateless, but he had an enormous curiosity about the world.

Einstein immersed himself in self-study to get into the Polytechnic Institute in Zurich, where he hoped to be able to study physics at a more advanced level. But he failed the entrance exam, and instead spent a year at a secondary school in Aarau, close to the Swiss Alps. Einstein was happy at the school, which placed great emphasis on developing the students' imaginative powers, rather than simply rattling off facts and figures from memory, and encouraged pupils to develop their visual thinking. He would come to make use of these skills – imagining and visualizing – in solving the puzzle of the absolute speed of light. A biologist can study an insect by holding it between their thumb and finger and observing it through a magnifying glass, but a physicist cannot pick up a ray of light to see what it is made of. That's why Einstein decided to study the characteristics of light with his imagination. He tried to visualize what he would see if he were running alongside a ray of light. If light consisted of electrical and magnetic waves, it ought to be possible to run alongside it so that the waves appeared to be stationary, just as a wave in the water might appear stationary to someone travelling alongside it in a boat at exactly the same speed. If the speed of the boat and the wave are exactly the same, the wave is completely still in relation to the boat.

But Einstein realized that the idea of running alongside a ray of light led to a paradox. No one had ever observed electromagnetic waves at a standstill, and according to Maxwell's equations, they couldn't possibly exist. It seemed impossible to conceive of what travelling alongside a ray of light would be like. Light waves must therefore operate differently from waves in the sea. To understand how this might be possible, Einstein had to call into question one of our most fundamental experiences of the world: the notion that space and time are the same for everyone. But it would be another ten years before he got there.

DAVID HUME SHOWS THE WAY

After his preparatory studies in Aarau, Einstein at last passed the Polytechnic entrance exam. But he was soon disillusioned, finding the teaching old-fashioned. Instead of listening to the lectures, he spent his time discussing philosophy and physics with his friends in the cafés of Zurich.

These discussions planted an important intellectual seed in Einstein's mind. He began to ponder the nature of time. Time is a constant presence in our lives – we are always talking about it, organizing our lives around it, and wishing we had more of it. But the question of what time really is has been thrown about by countless thinkers through the centuries. Einstein was particularly inspired by the Scottish philosopher David Hume. In

THE ASTRONOMER BY THE 'MOUNTAIN OF DEATH' 55

his book *A Treatise of Human Nature*, from 1740, Hume wrote that just as people form an understanding of space from the tangible objects in it, so too do we create an idea of time from the 'ideas and impressions' that succeed one another in our consciousness.[10] Hume wrote that it is therefore impossible 'for time alone ever to make its appearance'. Hume's assertion had a profound influence on Einstein. The passage of time is something we relate to by observing phenomena: the hands of a clock, the passage of the Sun across the sky, or, more abstractly, the flow of thoughts through our minds. Einstein realized that talking about time, and even space, becomes meaningful only in relation to how we measure them.

When Einstein later saw through our everyday understanding of time, he was able to solve the problem of the absolute speed of light. By that point, he had graduated from the Polytechnic, but he failed to make his way in academia. During his studies, several of Einstein's teachers had been irritated by his nonchalant, rebellious attitude. 'He always does something different from what I have ordered,' one teacher complained, while Einstein's professor told him, 'You have one great fault: you'll never let yourself be told anything.'[11] The professors at the Polytechnic refused to recommend him for potential jobs. Instead of a career in the academy, he was forced to lurch from one short-term private teaching job to the next.

When his partner and former fellow student Mileva Marić became pregnant, Einstein could no longer continue with his bohemian lifestyle.[12] He grew increasingly desperate to find a job. 'My scientific goals and my personal vanity will not prevent me from accepting even the most subordinate position,' he wrote to Marić.[13] When a friend told him about a position at the Patent Office in Bern, Einstein applied – and got the job. One of the most important physicists in the history of the world thus started his career as a civil servant, rather than as an academic, though Einstein didn't seem bothered by his new role. He wrote to a friend that an 'academic career in which a person is forced to produce scientific writings in great amounts creates a danger of intellectual superficiality'.[14] Perhaps he was right, because it was while working at the Patent Office that he revolutionized our understanding of the very nature of time and space.

THE INSTITUTE FOR PHYSICS

In the summer of 1902 Einstein started his new role as a patent clerk third class. His new job at the Patent Office was to evaluate applications to see if they followed patent law, and to rewrite them to ensure they were legally and technically correct. He worked eight hours a day, six days a week, and became so good at his job that he was able to spend part of every working day thinking about physics. He jotted down his scientific reflections on small scraps of paper. 'Whenever anybody would come by,' he said later, 'I would cram my notes into my desk drawer and pretend to work on my office work.' He called his desk drawer 'The Institute for Physics'.[15]

Later in life he wrote that it was at the Patent Office that 'I hatched my most beautiful ideas'. 'Pretending to work' gave him the time needed to ponder the properties of space, time and light without any requirement to get instant results. He realized that the reason the speed of light was so special might be found by questioning the very definition of what speed is. Take, for example, a person running a hundred metres. Assume, for the sake of argument, that they run the whole distance at the same speed. If we can measure the time it takes them to run that distance, we can find their speed. If it takes ten seconds, the speed is

THE ASTRONOMER BY THE 'MOUNTAIN OF DEATH'

ten metres per second. To take this measurement, we must use a tape measure and mark out a one-hundred-metre stretch. We also need a timer to find out how long it takes the person to run it. So, measuring a speed requires tape measures and clocks, or, to put it more abstractly, a measurement of intervals in space and time. Einstein began to suspect that the absolute speed of light could therefore give us an insight into the very nature of space and time.[16]

And indeed, what are space and time? Space seems to have three dimensions, because there are three possible directions we can move in. In relation to our bodies, these directions are up and down, right and left, and backwards and forwards. Our spontaneous conception of space is therefore as a kind of unchanging, static arena in which all objects exist.

While space is often something we take for granted, time is something we constantly wrestle with. It is slippery and elusive, yet tangible and concrete. We constantly come up against time in the form of clocks. They might assume a range of different forms, such as sundials, hourglasses, mechanical clocks and, in our day and age, numerals on an electronic screen. Even the sky provides a clock that informs us of the passage of time in the form of the movements of the Moon, the Sun and the stars.

And yet time is not the same as space. We can move back and forth through the three dimensions of space. Time's only dimension has but one direction: forwards (it's interesting how often we use spatial concepts to describe temporal phenomena). Time that has passed will never come back, while we can, in principle, always return to space we have moved through.

But even if space and time operate differently, we often imagine that they have one thing in common: they are the same for everyone. If you and I move from one point to another along the same stretch, but at different speeds, we should still cover the same distance. If our clocks are synchronized, a second on my clock should always be equivalent to a second on yours, regardless of how we are travelling. While working as a clerk at the Patent Office, Einstein realized that this assumption was wrong.

What if, Einstein reasoned, our experience of space and time depends on our speed? Let's return to our example of the person running a hundred metres in ten seconds. Let's make the situation a little more concrete: I run a hundred metres and you stand at the finish line and measure how long it takes. I too am measuring the interval with a watch on my wrist. When I've finished running, we compare how long it took. Assume you measure it as ten seconds (one of the liberties I can take as an author is imagining I can run a hundred metres in ten seconds, which unfortunately is not the case). But my watch says it took 9.999999999999994 seconds. A shorter time! We can assume that the clocks are precise enough that this isn't due to an error in measurement (no such accurate watch exists, but that doesn't matter, since this is merely a thought experiment). How could the two clocks give different times, even though they have measured the same interval?

Einstein's answer was that space and time are *not* static arenas that are the same for everyone. Space and time are relative! They change depending on our speed. The faster you move, the more your measurements of space intervals and time intervals will differ from someone who is standing still. This was the solution to the problem of the absolute speed of light. Einstein realized that when we move, time and space change in such a way that the speed of light will always be the same for all observers, regardless of their own speed.

The average human walking speed is around five kilometres an hour, while the speed of light is a little over a billion kilometres an hour. Day to day we don't notice the relativity of time and space because we move about at such negligible speeds compared to light. But say you want to go to Proxima Centauri. You decide

THE ASTRONOMER BY THE 'MOUNTAIN OF DEATH'   **59**

to travel there at 80 per cent of the speed of light (ignoring the insurmountable technical challenge of accelerating and decelerating at such speeds). When you reach the star you stop, take a few nice selfies with the star and then head back to Earth, once again at 80 per cent light speed. A friend is waiting for you on Earth. She's been measuring how long you've been away: just over ten years. And yet your clock says it only took you six years. Your speed has meant that you've aged more slowly than your friend who stayed on Earth.[17]

THE SPECIAL THEORY OF RELATIVITY

On 30 June 1905 Einstein published an era-defining article in which he presented his new ideas. The article can justifiably be said to be one of the most important in the history of physics. Einstein summarized his new insights into the properties of space and time in his special theory of relativity. 1905 is known as Einstein's *annus mirabilis* – his miracle year – since he published a total of four articles that laid the groundwork for the theory of relativity, the quantum mechanical description of light and the proof of the atom's existence – all while working at the Patent Office!

The special theory of relativity is one of the cornerstones of our understanding of nature. A common, yet mistaken, assertion is that Einstein's theory means that 'everything is relative'. That is not the case. Einstein meant something much more specific, namely that the distances and times we measure are relative in the sense that they depend on how we move. In the same way that a speed count is only meaningful in relation to something that seems to be 'resting' (the Earth's surface, for instance, or the Sun, or the galactic centre), distances and times are only meaningful if we know about the movement of the person doing the measuring. The speed of light, however, is absolute – that is, it is independent of our movement. Einstein's vision required a reassessment of which physical phenomena are absolute (the same for everyone) and which are relative (dependent on the movement of the observer). Space, time and simultaneity had been seen as absolutes, but were, according to Einstein, relative. The speed of light had been thought of as relative, but was, according to Einstein, absolute. In Newton's theory there was no limit to how high speeds could be, but according to the special theory of relativity there was a limit: in a vacuum, nothing can move faster than light.

There is a significant consequence to the fact that the speed of light is an absolute limit. If nothing can travel faster than light in a vacuum, no information can travel from one place to another faster than the speed of light. The speed of light sets a causal speed limit for the universe. When Einstein explored the characteristics of light, he therefore gained an insight into the fundamental terms of cause and effect in nature. This will turn out to be crucial for our understanding of black holes, as we will find out later.

But why was it Einstein, of all people, who had these new insights? At the time there were several scientists thinking along similar lines, such as the French mathematician Henri Poincaré and the Dutch physicist Hendrik Lorentz. They puzzled over the characteristics of time and space and the relationship between light, electricity and magnetism. The fact that Einstein was the one who took the essential step in formulating this new vision of space and time was probably due partly to his personality. He had a unique ability to focus on theoretical questions. Even as a child he could be completely consumed by a book, fully focused on it and apparently unaware of everything going on around him. He was able to approach theoretical problems from a range of

THE ASTRONOMER BY THE 'MOUNTAIN OF DEATH' **61**

perspectives, thanks to his well-developed ability to think visually. What's more, Einstein had a distinctly anti-authoritarian character, was never satisfied with traditional explanations and had not followed a classic academic path. Einstein's biographer Jürgen Neffe expressed it neatly: 'His academic failure was ultimately transformed into the secret of his success.' Einstein himself put it like this:

> When I ask myself how it happened that I in particular discovered the relativity theory, it seems to lie in the following circumstance. The normal adult never bothers his head about spacetime problems. Everything there is to be thought about it, in his opinion, has already been done in early childhood. I, on the contrary, developed so slowly that I only began to wonder about space and time when I was already grown up. In consequence I probed deeper into the problem than an ordinary child would have done.[18]

But it wasn't only Einstein's character that was important. The circumstances in which he lived also affected his thinking. New technologies were changing humanity's relationship to time, space, light and information. We've already seen how the electric light illuminated society. This gave physicists a tangible reason to explore the relationship between light and electricity, which Einstein was able to do in his youth in his father's and uncle's company. Wireless telegraphy enabled people to send messages through the air at the speed of light. The Atlantic underwater cable transported electrical signals between continents. The locomotive and the ever-expanding railways connected different regions, and increased people's speed through space. This increased speed also required that various ways of keeping time – which had sometimes differed from one town or country to the

62   FACING INFINITY

next – were unified in a single global time system. This new time
system required clocks to be synchronized over large distances
with the aid of electric signals, and it was patent applications for
just such systems that landed on Einstein's desk at the Patent
Office. While he smoked cheap Swiss cigarettes and evaluated
applications with titles like 'Installation with Central Clock for
Indicating the Time Simultaneously in Several Places Separated
from One Another', Einstein took the opportunity to reflect on
the properties of time.[19]

Einstein was well aware of how these new technologies
changed people's way of thinking about the world. His scientific articles and popular science works often featured clocks,
mirrors, magnets, electric cables and lights, as well as railway
carriages, embankments and platforms. The theory of relativity
was born out of an encounter between – on one hand – these new
everyday technologies and – on the other – a creative impulse
to understand phenomena related to space and time that lay far
outside the everyday sphere of experience.

One of Einstein's maths teachers, Hermann Minkowski,
took on the task of recasting Einstein's physical insights into
mathematical form. Minkowski was surprised that it should
be his student Einstein who had thought up these new ideas,
as he'd thought him a 'lazy dog' who 'never bothered about
mathematics at all' during his years as a student. But Minkowski
realized that there was a deep mathematical structure behind
Einstein's new theory. It was possible to unite space and time in
a common mathematical framework, which Minkowski called
spacetime. How far something travels in spacetime is absolute,
but how that journey is divided into its 'space' part and its
'time' part is dependent on the movement of the observer. In
a famous lecture, Minkowski stated that 'Henceforth space by
itself, and time by itself, are doomed to fade away into mere

THE ASTRONOMER BY THE 'MOUNTAIN OF DEATH' 63

shadows, and only a kind of union of the two will preserve an independent reality.'[20]

It is in spacetime that all galaxies, stars, planets, people, animals, stones, plants, particles and other matter exist. But when Einstein continued his explorations, he discovered that spacetime can be brought to life, becoming a dynamic actor in the drama of the universe – and the most extreme example of this is a black hole.

Chapter 4

· · · · · · · · · · · · · ·

EINSTEIN AND
THE BLIND BEETLE

Understanding how nature works on a fundamental level: that's the dream of physicists everywhere. This dream is not just a question of being able to explain concrete phenomena such as why the Earth spins on its axis, why the sky is blue or why the stars shine. No, the most fundamental level is about describing the basic laws underpinning the natural order. These laws govern how things move, interact, come into being and cease to exist; they have given rise to our universe with its stars and planets, molecules and atoms, complex life forms and, ultimately, conscious beings who try to understand the world they are living in. Therefore, if you can understand the laws of nature, you can understand the universe itself.

Newton is said to have used a lovely metaphor to describe this ambition. In his search for knowledge, which made him one of the greatest scientists in history, he saw himself as a 'boy playing on the seashore, and diverting myself in now and then finding a smoother pebble or a prettier shell than ordinary, whilst the great ocean of truth lay all undiscovered before me.'[1] One of the extraordinarily pretty shells he'd found there by truth's great ocean was the law of universal gravitation, which describes the strength of the gravitational force between two bodies, such as the Sun and the Earth, or the Earth and the Moon. But despite his great discovery, Newton was dissatisfied. 'Hitherto,' he wrote, 'we have explained the phenomena of the heavens and of our sea by the power of gravity, but have not yet assigned the cause of this power.'[2]

It was Einstein who managed to find the cause, and it was his discovery, which has come to be known as the general theory of relativity, that led to the assertion that black holes really can exist in the universe.

EINSTEIN'S HAPPIEST THOUGHT

'I am a respectable federal ink pisser with a decent salary,' Einstein wrote to a friend in the spring of 1907.[3] Despite having published four groundbreaking articles two years previously (one of which would lead to him being awarded the Nobel Prize in Physics in 1921), he had not yet managed to get an academic position. His only career progression was a promotion from patent clerk third class to patent clerk second class, and it was while he held this post that he pondered the great mysteries of the natural world.

Einstein realized that there was a problem with Newton's law of gravitation: it was not compatible with the special theory of relativity that he had presented in 1905. According to Newton's theory, the effect of the gravitational force is instantaneous. If the Sun suddenly disappeared, the Earth would immediately start travelling on a straight course out of the solar system. Day wouldn't become night, however, until after around eight minutes, since this is how long it takes light to reach the Earth from the Sun. Einstein thought that this was impossible. According to the special theory of relativity, no effect can occur faster than the speed of light in a vacuum. Therefore, it ought to take just as long for the Earth's course to change as it would for the last rays of light to reach its surface.

Einstein wanted to find a new, better theory of gravity that did not contradict the special theory of relativity. He started by analysing a well-known phenomenon that Newton had struggled to explain. It's an easy phenomenon to study. Take two objects of differing mass – a thick book and a thin one will work fine. Hold both objects at the same height above the ground and let go of both of them at the same time. It wouldn't be unreasonable to expect the thicker, heavier book to hit the ground first. But if you carry out the experiment, you'll find that both objects fall and hit the ground at the same time. Galileo himself had noted that, as long as air resistance does not influence the fall, all objects will land simultaneously if dropped from the same height.

In order to explain why objects with different masses fall at the same rate, Einstein needed to re-evaluate his understanding of the properties of space and time. In 1907 he made a decisive breakthrough, which he described, when he was older, as the 'happiest thought' of his life. His account of how it occurred has been quoted countless times, and it goes like this: 'I was sitting on a chair in my patent office in Bern. Suddenly a thought struck me: If a man falls freely, he would not feel his weight.'[4]

Einstein's description is pretty funny. He's sitting on a chair when he has this thought. Someone sitting on a chair can feel the weight of their own body against the chair. But what Einstein realized was that if someone is sitting on a chair that's in free fall through the air, they wouldn't be able to feel their body's weight in that chair (allowing for the proviso that this only applies if we disregard air resistance). This insight enabled Einstein to solve the mystery of gravity.

EINSTEIN'S SPACE LABORATORY

Let's take a look at the practical implications of Einstein's 'happiest thought' by following him on a thought experiment. Assume we have two laboratories out in space where we are undisturbed by the Earth's atmosphere. One is orbiting the Earth, while the other is on a straight course through the intergalactic void, far away from the Milky Way. Our laboratories have neither windows to look out of nor rockets to change their course. One is affected by Earth's gravity, while the other experiences no gravitational effect at all (hence the straight course it is travelling on – this happens when an object is moving with no external force acting upon it).

Now assume that you wake up in one of these laboratories without knowing which one it is. Einstein proposed that there is no experiment you can conduct that would tell you which of them you find yourself in. You can measure the passage of time, throw objects in different directions and beam lasers every which way, but none of these experiments would help you determine whether you're in the laboratory orbiting Earth or the one far off in outer space.

If you were to take an apple, for example, and place it beside you, it would continue to float in the same place regardless of which laboratory you were in. The laboratory orbiting the Earth is in a state of constant free fall. The apple would thus be falling just as you would be, since all objects fall in an identical manner under the influence of gravity, regardless of their mass. If you were in the laboratory in outer space, the apple would likewise continue to float beside you, since there is no gravity there to influence the apple's motion. Any experiment you might carry out would give the same result in both laboratories.[5]

This thought experiment led Einstein to a decisive finding. If you are out in the depths of space, free from all gravitational influence, you will follow the most fundamental course: a straight line. But if this situation is experimentally identical to being in free fall around the Earth, it follows that even there you must be moving in a straight line. Remember, there was no experiment that would enable you to distinguish between these two situations. Now, if we were to observe these two situations from the outside, they would be anything but identical. In one example you are travelling on a circular course around the Earth, in the other along a straight line. In one example you can feel the Earth's gravity, in the other you feel no gravity at all. From this similarity between two apparently different situations, Einstein came to an incredible conclusion: *when you travel in a circular orbit around the Earth, you are really moving along a straight line in curved spacetime.*

Gravity, Einstein thought, is about the geometric characteristics of spacetime itself. Instead of Newton's description of gravity as a force, Einstein introduced a new vision in which gravity was about geometry. In his special theory of relativity, Einstein had shown that spacetime is the arena upon which all the world's events play out. Now he realized that this arena could be twisted and distorted, or, in mathematical terms, curved. Free fall is straight motion in curved spacetime.

But what does spacetime being curved actually mean? Einstein spent many years searching for an exhaustive answer to this question.

ASTRAL, IMAGINARY AND NON-EUCLIDEAN GEOMETRY

'Grossmann, you've got to help me or I will go crazy.'[6]

By the summer of 1912 Einstein was desperate. He sought the support of his friend Marcel Grossmann, because he himself was incapable of dressing up his physicist's insights in mathematical

EINSTEIN AND THE BLIND BEETLE

clothing. Einstein knew that the solution to the mystery of gravity lay in understanding the geometric properties of spacetime. These properties describe, for example, the distance between two points, the sum of the angles in a triangle, or how fast clocks tick in different places. Einstein postulated that *all* forms of matter and energy alter space and time. But exactly how much, and in what manner? He needed to write down a formula that could state this in detail. He also needed a mathematical description of how different objects – from rocks on the Earth to planets in the solar system – might move in this mutable spacetime.

But no one had ever done this before. Visualizing spacetime is hard, and visualizing spacetime as curved is even harder. Einstein said, 'No man can visualize four dimensions, except mathematically... I think in four dimensions, but only abstractly.'[7] When visual thinking is found lacking, mathematics can step in, and Einstein needed to orient himself in a strange new geometric world. To do so, he needed help.

A common yet mistaken perception is that Einstein was an isolated genius. In fact, he commonly collaborated on problems with friends and colleagues, and when it came to the geometry of spacetime, he contacted his old friend Marcel Grossmann. Grossmann had helped the bohemian Einstein with mathematical questions while he was studying and had tipped him off about the job at the Patent Office. Grossmann became a mathematics professor at the Swiss Federal Institute of Technology (known as the Federal Polytechnic until 1911), where Einstein had also at last managed to get a position (he sarcastically said he'd become 'an official member of the guild of whores').[8] Now Grossmann would once again assist Einstein at a crucial stage.

In order to succeed, Einstein and Grossmann needed to get deep into non-Euclidean geometry, which, as the name suggests, contains features that are fundamentally different from Euclidean geometry. Geometry means 'measuring the Earth'. An ancient Babylonian clay tablet, dated at around 3,700 years old, provides a clue as to the origins of geometry. The clay tablet describes a dispute in Babylon between a person called Sin-bel-apli and a wealthy landowner.[9] They are unable to agree on whose land a number of valuable date palms stand, and a surveyor is called upon to calculate where the boundary between the disputing parties' land lies. In order to do this, the surveyor has to measure angles and distances, and then use these observations to calculate the extent of Sin-bel-apli's land. Geometry thus originated in attempts to calculate the size of plots of land. It was a practical skill, equivalent to fishing or carpentry, that was used in questions around land ownership, architecture and agriculture.

Then, in the third century BCE, the Greek mathematician Euclid elevated geometry from its commonplace status into a branch of abstract mathematics.[10] Although he is one of history's most important mathematicians, we know little about him; details of his birthplace and appearance have faded into obscurity. But Euclid's contributions live on in the form of one of the best-known books of all time: *The Elements*. Euclid's goal was a systematic one: he wanted to gather and present the shared mathematical knowledge of his time as cogently as possible. Even today, *The Elements* is held up as an example of how to construct logical reasoning through definitions, postulates and derivations. Einstein himself was captivated by the book when he read it as a boy.

But there was a problem with Euclid's geometric system. He had sought to establish a series of postulates from which to derive geometric relationships for surfaces, triangles, circles and so on. These postulates were to be intuitive and easy to understand, such as, 'A straight line segment can be drawn joining any two points', or 'All right angles are equal to one another'. But the

EINSTEIN AND THE BLIND BEETLE

fifth postulate was anything but intuitive. It states: 'If two lines are drawn that intersect a third in such a way that the sum of the inner angles on one side is less than two right angles, then the two lines inevitably must intersect each other on that side if extended far enough.'

I had to read that convoluted sentence several times before I finally understood what Euclid meant: according to the fifth postulate, two parallel lines will never meet or diverge. If you were to draw two such lines and imagine them stretching to infinity, the distance between them ought to stay constant, so they can never cross or move away from one another.

Euclid thought this was completely reasonable, but he was disappointed with the complexity of his postulate. It should be possible to derive it from the other postulates, but neither Euclid nor any other mathematician succeeded. The fifth postulate therefore became a source of frustration in this otherwise elegant and orderly geometric system. The Persian poet and mathematician Omar Khayyam formulated this frustration powerfully in 1077: 'How can one allow Euclid to postulate this proposition... while he has demonstrated many things much easier than these?'[11]

Few doubted that parallel lines have the properties Euclid proposed; the very sign for 'equals' represents two lines that run in parallel: =. The Welsh mathematician Robert Recorde introduced the symbol in 1557 because he was of the opinion that 'noe 2 thynges can be moare equalle' than two parallel lines.[12]

In spite of several attempts over more than two millennia, no mathematician managed to solve the problem of Euclid's fifth postulate. It seemed so simple and trivial – parallel lines never meet – but was still so hard to prove. But on 3 November 1823 the Hungarian mathematician, fencer and dancer János Bolyai wrote to his father Farkas, 'I have discovered such magnificent things that I am myself astonished at them. It would be damage eternal if they were lost.'

János had discovered non-Euclidean geometry. Just like his father, János had become fascinated by the mystery of Euclid's fifth postulate. But his father, who was a noted mathematician and author, had warned his son against wasting time on this geometric conundrum. Farkas Bolyai said that, in studying the fifth postulate, he had 'traversed this bottomless night, which extinguished all light and joy in my life'. Just like other mathematicians before him, Farkas Bolyai had tried in vain to prove that parallel lines would never cross or diverge from another. But he had failed, and therefore he exclaimed to his son, 'For God's sake! I entreat you, leave the science of parallels alone.'[13]

Like thousands of sons before and since him, though, János refused to listen to his father's advice. The ambitious son not only wanted to be the best fencer and dancer in the Imperial Austrian Army, he also wanted to solve the problem that had hitherto evaded mathematicians. Like his father, he set his sights on solving Euclid's problem – and succeeded. Instead of trying to show how Euclid's fifth postulate follows from the other postulates, János Bolyai started from the opposite direction. What happens if parallel lines actually do diverge? A whole new world of geometry opened up for him. In this world, the sum of the angles of a triangle did not need to be 180 degrees, Pythagoras's theorem ceased to apply, the relationship between the circumference of a circle and its diameter did not need to equal π and, above all, the distance between parallel lines did not have to be constant. 'Now I cannot say more, [except] that from nothing I have created a wholly new world,' János exclaimed when he analysed these strange geometric properties.

Farkas Bolyai was astonished by his son's discovery. He encouraged him to publish his results as soon as possible. Farkas was

EINSTEIN AND THE BLIND BEETLE

concerned that other mathematicians would arrive at the same conclusions, as '[w]hen the time is ripe for certain things, these things appear in different places in the manner of violets coming to light in early spring'. He was right. János wasn't the first. The Russian mathematician Nikolai Lobachevsky published the article 'A Concise Outline of the Foundations of Geometry' in the journal *The Kazan Messenger* in 1829. Lobachevsky was a professor, and later the rector of the university in Kazan, the capital of the present republic of Tatarstan, east of Moscow. Unbeknown to Lobachevsky, János Bolyai published his book *The Science Absolute of Space* in 1831. Both had made similar discoveries, but neither knew of the other's existence, working, as they were, on the margins of the scientific world. Ironically, Carl Friedrich Gauss – who worked in Göttingen and is one of history's foremost mathematicians – had come to similar conclusions several years previously. He, however, had chosen not to publish his results, perhaps because he feared the world was not yet ready to appreciate these new geometric realms.

Initially, this new geometry went under various names: imaginary geometry, astral geometry, anti-Euclidean geometry, pangeometry, but in the end the name non-Euclidean geometry stuck. It is a kind of geometry that differs from our everyday understanding of geometric relationships. Our practical daily requirements and our limited experience of the world have led us to develop a kind of geometric intuition of how distance, area, volume and angles work. We're used to thinking in Euclidean terms, because this way of thinking applies to geometric relationships on flat surfaces. After all, the Earth seems flat provided we are looking at a small enough region. Across larger areas of the Earth's curved surface, however, the sum of the angles of a triangle does not need to be equal to 180 degrees. But the presence of curved surfaces in space that have non-Euclidean geometry is one thing; the question of whether space itself has a non-Euclidean structure is much more fundamental.

Breaking with our everyday intuitions about the geometry of space required a mathematical bravery and a liberation from preconceived ideas about the nature of the world that oddly enough crystallized simultaneously in Lobachevsky, Bolyai and Gauss. It was one of the greatest discoveries in the history of mathematics, yet it was received, not with fanfares, but with deafening silence. Neither Lobachevsky nor Bolyai received recognition for their discoveries while they were alive. Lobachevsky died blind and destitute. Bolyai sustained a serious head injury in a horse-riding incident, never fully recovered, and died in obscurity. The great Gauss, who had kept his results secret from the world at large, put his student Bernhard Riemann to work on developing non-Euclidean geometry further. Riemann was a brilliant mathematician, but he was sickly, reclusive and poor. He extended Gauss's insights to more spatial dimensions and with more varied geometries than those Lobachevsky and Bolyai had discovered, before dying of tuberculosis at a relatively young age. The pioneers of non-Euclidean geometry did not have an easy time of it.

The key finding in the work of Lobachevsky, Bolyai, Gauss and Riemann – one that would be crucial to Einstein and later to our knowledge of black holes – is that the geometric characteristics of space must be determined by experimental means. 'All mathematical principles which are formed by our mind independently of the external world will remain useless,' wrote Lobachevsky. Whether the sum of the angles in a triangle is 180 degrees or not is something that must be determined by measurement. Lobachevsky tried to measure the angles formed by the triangle connecting the Earth, the Sun and Sirius, the brightest star in the night sky that today is known to lie 8.6 light years

EINSTEIN AND THE BLIND BEETLE

away. Lobachevsky could not find any deviation from Euclidean geometry in the celestial configuration, and concluded that if the universe had a non-Euclidean structure it had to operate on distances much larger than those to Sirius. Gauss tried to make similar measurements too. In the course of an extensive geodetic survey of the Kingdom of Hanover, he travelled round in rickety carts and measured distances and angles between towers and hills. He took the opportunity to measure the sum of the angles between Brocken, Inselberg and Hoher Hagen. The three mountains form a triangle, and Gauss was able to ascertain, with an accuracy up to 0.0002 per cent that the sum of the angles was 180 degrees. Space was Euclidean, at least in central Germany.

'The importance of C.F. Gauss for the development of modern physical theory and especially for the mathematical fundament of the theory of relativity is overwhelming indeed,' Einstein wrote.[14] But the influence of non-Euclidean geometry stretched beyond science. The new geometry found its way into art and literature too. The Oxford mathematician Charles Lutwidge Dodgson was inspired by it when he wrote (as Lewis Carroll) *Alice in Wonderland*, the horror writer H.P. Lovecraft sent his characters loose in its worlds, and the visual artist M.C. Escher created a whole oeuvre through his explorations of its peculiar visual consequences.

SEARCHING IN THE DARK

With the help of non-Euclidean geometry, Einstein was able to describe the curvature of space. But his goal was to understand the characteristics of spacetime as a whole. He therefore needed to consider the curvature of time, namely the way clocks tick at different rates in different places – not because they're out of sync, but because the passage of time is not the same everywhere.

Working with Grossmann, Einstein began to write down equations describing the links between matter and energy on one hand and the geometry of spacetime on the other. The work of finding the correct formula was a long, arduous trek through the mathematical wilderness. The equations he and Grossmann came up with were insufficient. They were either incapable of accounting for what we actually know about how gravity works in our solar system, or they suffered from mathematical shortcomings (which Einstein called 'a hair in the soup').[15] He complained to friends that 'each step is devilishly difficult' and that it took 'years of searching in the dark for a truth one feels but cannot express'.

Their woes were compounded by the First World War. By the time it broke out Einstein had established himself as a professor in Berlin. He despised the war. 'I'd rather be beaten black and blue than take part in such a wretched business!' he wrote about the military's activities.[16] In a letter to the French philosopher and pacifist Romain Rolland he summarized his view on the war: 'A decisive victory for Germany would be a misfortune to the whole of Europe.'[17]

In spite of the war, Einstein continued to search for the right formula to describe how matter and energy change space and time. He called it 'eavesdropping on nature'.[18] In the autumn of 1915 he was working extra hard. The mathematician David Hilbert in Göttingen was also trying to find the right formula, and Einstein was drawn into a fraught battle over who would get there first. He made frantic calculations, forgot to eat, smoked frenetically, slept irregularly; his efforts were repaid only with stomach problems. He had unified space and time, but he was watching the world around him falling apart. His marriage broke down,

EINSTEIN AND THE BLIND BEETLE

he was barely able to see his son, his colleagues were mired in bitter nationalism, and death was spreading like wildfire through a war-torn Europe.

In the end Einstein managed to find the right formula, but the search had taken its toll. He confided in a friend that he had lived through 'the most exciting and demanding time of my life'.[19] On 25 November 1915 the thirty-six-year-old professor ventured out of his apartment in the south-west of Berlin and made his way to the Prussian Academy of Sciences. The city streets bore the stamp of war and starvation. Women crowded outside steaming soup kitchens to get just a little food for their families. There was an acute food shortage due to the British Navy's blockade of food and agricultural deliveries via the North Sea. Black market prices for meat, lard and butter were sky-high. The police had orders to fire warning shots to dissipate the increasingly frequent hunger riots.

Einstein passed the magnificent entranceway of the Prussian State Library on Berlin's grand boulevard Unter den Linden, where the Academy of Sciences held its meetings. Now he would finally present to his colleagues the result of ten years' struggle: a formula that provided a new vision not only of how gravity worked, but of the very nature of space and time. Einstein had written down ten equations that described how space and time curved in the presence of matter and energy. The equations are known collectively as the general theory of relativity.[20] The basic principle of this theory is that all matter and energy in the universe change the geometry of spacetime. In turn, this geometry determines how everything moves.

On Earth, gravity relies chiefly on the curvature of time. If we were to stack synchronized clocks on top of one another all the way out into space, they would stop ticking at the same rate. The lowest clock would go slower than the one above it, and so on.

The higher the clocks get, the faster they go. The reason for this is that the mass and energy of the Earth change how fast time passes. The characteristics of space change too, but this has significantly less effect than the altered passage of time.[21] The primary reason that objects fall towards the ground is therefore, according to Einstein, that time passes more slowly there. That is the cause Newton failed to find. The US physicist and Nobel laureate Kip Thorne, who we will encounter several times in this book, has a poetic way of expressing the true cause of gravity: 'Everything likes to live where it will age the most slowly, and gravity pulls it there.'[22]

When Einstein formulated the general theory of relativity, no one had measured how much more slowly time passes close to the ground than it does higher in the air. Today this is a routine undertaking. In 2010 scientists at the National Institute of Standards and Technology (NIST) in the USA issued a press release entitled 'NIST Clock Experiment Demonstrates That Your Head Is Older Than Your Feet'.[23] Scientists had measured how a clock placed one-third of a meter above another clock ticked at a faster rate. Over a human lifespan of almost eighty years the time difference is very small, only ninety billionths of a second. In order to measure this difference, NIST scientists had to use atomic clocks so precise they lose or gain no more than a second every 3.7 billion years.

But the most obvious evidence of the effect of time's relativity on our everyday lives is our mobile phones. When we use navigation apps to find our way to places, our phones read space and time coordinates from a series of navigation satellites. The most famous example is the Global Positioning System, also known as GPS. The satellites race around, 30,000 metres above the Earth's surface, at a speed of more than fourteen thousand kilometres an hour. The speed of the satellites causes their clocks

EINSTEIN AND THE BLIND BEETLE

to run slow relative to clocks on the Earth's surface, while the difference in the gravitational field between the surface and the satellites' position causes the clocks on the satellites to run fast. The combined effect is that the clocks aboard the satellites run fast by thirty-eight microseconds every twenty-four hours compared to clocks on the surface. Thirty-eight microseconds might not sound like much, but if the algorithms that calculate our location did not take account of this time difference, the navigation system would soon become unusable. The functioning of our global economy and our ability to accurately navigate the Earth's surface is therefore dependent on Einstein's relativity theory.

THE BLIND BEETLE

It was Einstein's analysis of the speed of light that gave rise to the special theory of relativity, and his analysis of why bodies of different mass fall at the same rate that led to the general theory of relativity. With special relativity he had united space and time into spacetime. With general relativity he brought spacetime to life. Spacetime became not only the arena upon which our lives play out, but an actor that could be altered, bent, stretched and squeezed.

There are real lessons to be learned from Einstein's path towards the theories of relativity. He is said to have explained the reason he had become so famous to his son by saying that, 'When a blind beetle crawls over the surface of a curved branch, it doesn't notice that the track it has covered is indeed curved. I was lucky enough to notice what the beetle didn't notice.'[24]

The beetle represents humanity, and the branch represents space and time. As we go about our daily lives we do not become immediately aware of the curvature of space and time, but through an enormous mental effort, Einstein was able to unravel their dynamic nature. When talking to his son he called it 'luck', but in reality Einstein engaged, for ten years, in a mental struggle to go beyond our everyday understanding of the world and see it from a new perspective.

Einstein's story of the blind beetle shows how important it is to nurture and retain our curiosity and imagination. Without these attributes, he would never have been able to develop his new theories. Another lesson we should take from this is to be wary of jumping to conclusions about how the world works on the basis of our limited everyday experience. This applies not only to Einstein's insights into the relativity of space and time. Quantum mechanics, another fundamental theory from the early decades of the twentieth century, and one that Einstein contributed to, also has many elements that are hard to reconcile with our day-to-day experience of the world.

AN ANOMALY IN THE SOLAR SYSTEM

The first thing Einstein did to check whether his new theory worked – that is, whether its predictions aligned with what physicists and astronomers had observed – was to investigate what it predicted with regard to the orbit of Mercury. It takes eighty-eight days for Mercury to orbit the Sun. When the planet has completed one orbital period, however, it does not return to exactly the same spot. This is due to the fact that, as well as the Sun, the other planets in the solar system exert a gravitational influence on Mercury, so the course of its orbit shifts slightly each time.

In the middle of the nineteenth century, the French astronomer Urbain Le Verrier realized that Newton's theory did not give

EINSTEIN AND THE BLIND BEETLE

an accurate figure for the shift in Mercury's orbit. The planet's perihelion – the point on its orbit closest to the Sun – seemed to shift in a way that was not explained solely by the effect of the solar system's other planets on Mercury's course. A possible explanation for this anomaly was that another planet was contributing to the shift. Since this had to be within Mercury's orbit, which would mean it was heated intensely by the Sun, this hypothetical planet was given the name Vulcan. Several astronomers claimed to have seen Vulcan, but these observations were never confirmed.[25]

Einstein tried to calculate what his new theory might tell us about Mercury's orbit. However, his equations are complex: if you were to write out all the mathematical terms contained within them, they would occupy many pages. Einstein therefore felt compelled to simplify his own equations, and when he applied these simplified versions of his theory to Mercury's orbit, the result set his heart racing. The equations predicted that Mercury would move exactly as the astronomers had observed. It was the first proof that Einstein's theory gave the right answer, where Newton's theory fell short.[26]

Einstein wrote to a friend that he was 'beside himself with joy'. He presented his calculations at a meeting of the Academy of Sciences in 1915. Karl Schwarzschild was probably there in the audience, listening to Einstein's presentation.[27] When Schwarzschild returned to the Western Front he immediately set to solving Einstein's equations without simplifications. He succeeded after just a few weeks, and it was then that black holes appeared – in the form of a mathematical spacetime formula.

Chapter 5

· · · · · · · · · · · · · ·

BEYOND THE
EVENT HORIZON

'I 've spent three days doing mindless ballistics calculations,' Karl Schwarzschild wrote to his wife Else in the autumn of 1915.[1] As we've seen, he was stationed in Mulhouse, on the Western Front. After German troops had seized control of the town at the beginning of the war, they occupied a house that had previously belonged to a French textile manufacturer. Schwarzschild was living in one room of the house, which also had a dining room, a billiard room, an office and a piano room. He wrote to Else that his room was heated and had electric lights, that he spent the evenings playing billiards, and that the weather in Mulhouse was warm and wet.

This was the backdrop against which Schwarzschild hit upon the formula for a black hole. When not conducting his military calculations, Schwarzschild was immersed in solving the equations that made up Einstein's new general theory of relativity. He wanted to know what they could tell him about the movement of the planets and the interior of stars.

In the archive in Göttingen, I go through the folders containing Schwarzschild's archive. The spring sunlight that was filtering through the window has dimmed as a light drizzle has begun to fall over the town. In one folder I find the large sheets of paper on which Schwarzschild made his calculations.

His handwriting is hard to decipher, but I immediately recognize the mathematical symbols. They're the same symbols learned by all students of general relativity today, the ones I often used when I was doing research for my PhD.

Schwarzschild's mathematical notes highlight his struggle to understand how space and time change in and around a star. I picture Schwarzschild sitting at a desk in his room in Mulhouse. In the glow from an electric table lamp his pen moves over the paper. Beside him is the article in which Einstein described his new field equations for the general theory of relativity. Schwarzschild wants to explore the equation's applicability to the problem of Mercury's orbit without making the simplifications Einstein felt compelled to make. He performs the calculations successfully.

'Esteemed Mr Einstein,' he wrote in a letter posted at the end of 1915. 'In order to become versed in your gravitation theory, I have been occupying myself more closely with the problem you posed in the paper on Mercury's perihelion.'[2]

A thrill goes through me when I turn over Schwarzschild's letter: I've found what I was looking for. In his scrawled handwriting Schwarzschild has written the formula for a black hole. This was how the human search for these objects began in earnest – with a mathematical formula sent from the Western Front during the First World War. But neither Einstein nor Schwarzschild initially realized the implications of the formula. It didn't resemble anything physicists or astronomers were accustomed to studying. Instead of describing the properties of particles or ballistics, gases or liquids, it stipulates how space and time change around a star.

'I would not have thought that the strict treatment of the [mass-]point problem was so simple,' Einstein responded in a letter that's preserved in one of the folders.[3] Schwarzschild's formula does indeed describe the effect of a star's gravity on space and time. In the parlance of the general relativists, such a

84 FACING INFINITY

formula is called a spacetime metric, since it describes how space and time warp and twist. Taken to its extreme, in a hitherto unexplored mathematical world, the Schwarzschild metric not only shows the curvature of space and time around a star, but also how black holes work.

For more than fifty years, physicists and mathematicians sought to understand the strange mathematics at work in the Schwarzschild metric. It states that a black hole is a place in the universe where space and time have changed so much that nothing can travel away from it: not light, not matter, not information. Let's now take a look at what the formula predicted about a black hole's characteristics. If you're feeling uncertain about how a black hole works, you're in good company: most of the physicists studying them have felt this way at some point or other.

THE EVENT HORIZON

We'll begin our exploration of how black holes are formed by looking at a concrete example. As I read Schwarzschild's letters in the Göttingen archive, I'm sitting on a chair. I can feel the weight of my body against the chair. It's the mass of the Earth – its oceans, mountains, continents and inner matter – pulling on my body (to use a phrase that's more at home in Newton's conception of gravity). But the different parts of the Earth pull on me to different extents: a bit of matter on the other side of the planet exerts less pull on me than a similar bit right beneath my feet.

Let's carry out a thought experiment now, in which all the Earth's mass is compressed into a smaller volume. The continents would be deformed, the oceans would flood the land, and earthquakes and huge volcanic eruptions would break out all over the surface of the planet. But let's imagine that I'm still sitting on

BEYOND THE EVENT HORIZON

my chair, which is attached to the surface. I would feel, then, that my body was being pulled more strongly towards the Earth, because the distance between the different parts of the planet had decreased. The matter on the other side of the globe would now be closer to me, thus exerting a stronger gravitational pull. The more the Earth is squeezed together, and as long as its mass remains the same, the more the distance will decrease between my place on its surface and all the matter that it contains. This would increase the Earth's gravitational force on me, and make it harder for me to get up off my chair. If the Earth went on compressing, its gravity would eventually be so strong I'd be pushed down to the ground, unable to get up. If this process continued, not even light would be able to leave the Earth's surface. Our planet would have become a black hole. Its gravitational force would have become so strong it would go on collapsing in on itself.

This situation is distinct from the dark celestial bodies imagined by John Michell and Pierre-Simon Laplace at the end of the eighteenth century. In their conceptualization a black hole was a stable star that sent out light which then returned to its surface. But, according to Einstein's relativity theory, a black hole is not a material object that retains all light. When gravity becomes too strong, all matter collapses completely – what's left is nothing but a distortion of spacetime in the form of a black hole. So, what Schwarzschild had discovered was much stranger than the dark bodies that Newton's gravitational theory had led Michell and Laplace to take an interest in.

The surface of a black hole is called the event horizon. The term was coined by the Austrian-born physicist Wolfgang Rindler in 1956, who defined the event horizon as 'a frontier between things observable and things unobservable'.[4] No light, no information and no matter can pass the event horizon from the inside to the outside of the black hole. This is why the event horizon represents a boundary for what is observable: we cannot find out what happens inside the black hole without travelling into it.

The word 'horizon' calls to mind the line that separates the sky from the sea and the ground here on Earth. But an event horizon does not work in the same way. When the Sun sets below the Earth's horizon, we still see its rays lighting up the sky as it disappears. If we move towards the sunset, we can see even more of the Sun than we would if we'd stayed in the same place. On Earth, the horizon is relative: it depends on our position. But the black hole's event horizon is an absolute boundary that divides space into an inside and an outside. If a star were to fall beyond the event horizon, its light would disappear forever – in contrast to the Sun, which will rise again after it has set.

Neither is the event horizon a surface in the same way as the Earth's surface, which consists of matter. According to Schwarzschild's formula, a black hole does not consist of matter at all, but rather of space and time. In the Prologue we saw how the event horizon could be described using a metaphor: the river flowing towards a waterfall. The fastest pace at which a human can swim creates an invisible boundary in the water, beyond which no one can swim in the opposite direction. In a similar way, the event horizon is a boundary in spacetime that light cannot pass from the inside. Since the special theory of relativity states that nothing can travel faster than light, it is impossible to travel from the inside of the black hole to the outside. It's equally impossible to hover on the event horizon: once there, no one can avoid being dragged into the black hole. And if you're inside, space and time will be so altered that there will be no way out again. In that sense a black hole is the ultimate cosmic prison: no one can escape.

A black hole is not only dark because we cannot see what is happening inside. The wavelength of light travelling from the area close to the event horizon gets longer and longer, and its energy decreases as it gets further from the black hole. This is known as the light's *redshift*, because red light has the longest wavelength on the visible light spectrum. But the redshift applies to all electromagnetic radiation, from radio signals to gamma rays. The closer the light is to the event horizon when it begins its journey away from the black hole, the greater its redshift. Right on the event horizon the redshift becomes infinite as the light's wavelength is stretched to infinity and it loses all energy. That's why signals sent from just outside the event horizon are impossible to detect, and that's what makes black holes – according to the Schwarzschild metric – completely black.

In German, Schwarzschild means 'black shield'. The name derives from the Jewish quarter in Frankfurt, where Karl Schwarzschild's ancestor Liebmann Wohl lived in the 1500s.[5] Instead of having numbered addresses, some houses in the district had painted shields nailed to the exterior. The colour of the shield showed which family lived in which house. Black was the colour of the Schwarzschilds' shield, and so the family's address became 'zum schwarzen Schild', or 'by the black shield'. This eventually became the family's surname. Today that black shield from the Jewish quarter has given its name to one of the universe's darkest phenomena, via the Schwarzschild metric. The name is undeniably fitting, since the event horizon is a dark surface that, like a shield, prevents light from exiting the black hole.

THE SCHWARZSCHILD RADIUS

Using Schwarzschild's formula it is possible to calculate what volume an object with a given mass needs to shrink into to become a black hole. This volume tends to be given as a radius, known as the Schwarzschild radius. For a mass equivalent to a human body, the Schwarzschild radius is ridiculously small: just a billionth of the size of an atomic nucleus. The Earth's mass has a Schwarzschild radius of only eight millimetres. This means that its entire mass must be compressed into a sphere no more than sixteen millimetres in diameter in order to become a black hole. It's hard to imagine this kind of extraordinary density: Earth in its entirety squeezed to the size of a marble. The thought experiment we engaged in above, where the Earth became a black hole, is therefore nothing more than a fantasy – in reality it would be impossible.

For the Sun the Schwarzschild radius is around three kilometres. This means that for the Sun to become a black hole it would have to be squeezed into a volume with a diameter roughly equivalent to a city centre. That too is impossible, but in the next chapter we will see which cosmic processes can make stars with large masses implode and become black holes.

Even though the Sun cannot become a black hole, it is interesting to imagine what would happen to the planets' orbits if such a thing did occur. So, picture this: one beautiful spring morning, you and I are walking along the tops of some cliffs by the sea. We look up at the sky and feel the Sun's rays caress our faces. Then, all of a sudden, the sunlight dims. The air grows colder. Within the space of a few minutes, the Sun shrinks and its light goes out. The blue sky grows dark; we can see the stars but the Sun has gone completely black.

Our imaginary Sun-turned-black-hole is six kilometres in diameter. Will it swallow Mercury, Venus, Earth and the rest of the planets? No. A common, but mistaken, view is that black holes are like cosmic vacuum cleaners that suck in all matter and grow uncontrollably. But the Earth will continue to travel around our Sun-turned-black-hole in exactly the same way as before. The planets' orbits are unchanged, because, at a distance, a black hole's gravity works just the same as the gravity of a star of the same mass. It's only at close range that its peculiar warping of spacetime has an effect. In addition, the tiny size of the Sun-turned-black-hole in relation to the wider solar system makes it hard to navigate straight into it.

Black holes can come in a range of sizes. The distance from a black hole's centre to its event horizon is given by the Schwarzschild radius. This radius increases at the same rate as the mass of the black hole. How big a black hole is will thus be determined by how much mass it contains. As we have seen, the Schwarzschild radius for an object with the same mass as the Sun is three kilometres. If the mass were ten times larger, the Schwarzschild radius would also be ten times larger, i.e. thirty kilometres.

By following the orbits of the stars around Sagittarius A*, Andrea Ghez and Reinhard Genzel determined its mass to be more than four million Suns. In order to find out whether it was a black hole they used Schwarzschild's formula to calculate its size. If it were a black hole, it would have a radius seventeen times that of the Sun (corresponding to a size less than the orbit of Mercury). From the stars' orbits, Ghez and Genzel were also able to show that the mass of Sagittarius A* was concentrated into a very small volume, strong proof that it was indeed a black hole. Still, the Nobel committee thought it was open to discussion whether what we call a black hole is really equivalent to what the general theory of relativity predicts. Therefore the Nobel Prize was awarded to Ghez and Genzel for the discovery of a 'supermassive compact object', rather than a black hole. The committee concluded its scientific motivation for the Nobel Prize in Physics 2020 with the words, 'Nature may still have surprises in store.'[6]

FROZEN TIME

A clock on the surface of the Earth ticks more slowly than a clock higher up. In the previous chapter we saw that this effect was minimal on Earth, but close to a black hole the time difference can be much larger – even infinitely large.

The reason for this is the curvature of space and time. If you travel towards a black hole in a spaceship, you can head away at any point, risk-free, provided you stay outside the event horizon (assuming your rockets are powerful enough). The manoeuvre might take you a few hours, but for your friends waiting back at a space station, a whole lifetime might have passed. The time difference depends on how close to the black hole you come and what its mass is.

This possibility is predicted by the Schwarzschild metric, though it's hard to test this prediction experimentally since we can't travel to a black hole. That hasn't stopped science fiction films making great use of the relativity of time around a black hole.

In the 2014 film *Interstellar* (warning: spoilers ahead!), the crew of the spaceship Endurance have arrived at the gigantic black hole Gargantua, which is orbited by several planets. A previous expedition crashed on one of them and the Endurance crew want to find out what happened. One of the astronauts stays on the spaceship while the rest head down to the planet's surface. They know the mission is risky. Because the planet is close to the black hole, time passes more slowly there than on the Endurance and on Earth. The astronaut left on the spaceship warns the crew that every hour on the planet is equivalent to seven Earth years.[7]

Once on the planet, the crew are stranded because the motors on their lander fill with water. The viewer watching this scene can hear the clock ticking in the background: it's part of the music written for the film by the renowned composer Hans Zimmer. There are 1.25 seconds between each tick – equivalent to almost twenty-four hours on Earth.

In the end, the crew manage to return to the Endurance. It's not clear in the film exactly how long the mission took; it was probably only a few hours. In spite of this, the astronaut who was left on board has aged by more than twenty-three years, since time flows more slowly closer to the black hole than further away. The pilot of the Endurance learns via messages from Earth that his children have grown up while he was on the planet's surface. Much of what is portrayed in *Interstellar* is science fiction – but not the relativity of time around a black hole.

AT THE LIMIT OF KNOWLEDGE

The property of black holes that I find most interesting concerns one of the most fundamental aspects of nature: causality. Understanding how causality functions in nature is a key ambition of physics, since it describes how an event (the cause) at one point in space can influence another event (the effect) at the same location or at another point in space at a later juncture.

An event can have many different causes: an atom's radioactive decay, for instance, or you calling out 'hello' to someone you see on the street, or a plasma storm bursting out from the Sun's surface. An event can, in turn, create a result somewhere else.

Imagine that you say 'hello' to me from across the street. This makes air vibrate inside your mouth. The vibrations move through the air at a speed of around 340 metres per second. After a short while the vibrations reach my ear. I turn around and see it's you who called. We have two events here: you calling out 'hello' and me turning around. Your 'hello' is a cause, me turning around is an effect. Between this cause and effect the sound had to travel through the air, meaning it took a certain amount of time for your cause to produce my effect. The following formula describes this causality:

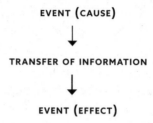

The transfer of information consists of some physical phenomenon – for example, sound or light – that can transport information from one point in space to another. This transfer can be fast or slow. As we saw in Chapter 3, the fastest transfer possible takes place at the speed of light in a vacuum. Because the special theory of relativity states that nothing can travel faster than this, the speed of light sets a limit for causal influence in nature. The result of this is that no external causal influence can occur from within the black hole. This makes the event horizon a causal boundary in spacetime and a limit upon our knowledge of what happens in nature. The physicist David Finkelstein mapped the mathematical structure of the event horizon in 1958, describing it as 'a perfect unidirectional membrane: causal influences can cross it but only in one direction'.[8]

Black holes can therefore be defined in several ways.[9] We can view them as a large concentration of matter, but it's also possible to see them as pure spacetime geometry. Furthermore, they act as a boundary for knowledge, causality and information transfer. One more way of defining them is to look at them as the end point of matter, that which remains at the point at which

BEYOND THE EVENT HORIZON 93

gravity dominates completely, making matter collapse in on itself endlessly – until the singularity.

THE END OF SPACE AND TIME

One of the biggest questions raised by black holes is what happens to things that fall into them. Because we can't study this from the outside we can only investigate the interior of a black hole with the aid of mathematics. Inside a black hole there is, according to the Schwarzschild metric, a point so infinitely dense that space and time themselves seem to cease to exist there. This point is known as the singularity.[10] Everything that enters a black hole will eventually reach it. This is not because the singularity is a point in space located at the centre of the black hole. Inside the black hole, space and time have been altered so much that our outsider's understanding of what ought to constitute a midpoint is no longer applicable. There is a complicated spacetime geometry at play there that changes over time, making the singularity a point in the future rather than in space. Because it is impossible to flee the future, it is impossible to avoid the singularity.

The term 'black hole' does a good job of capturing what the event horizon and the singularity are. The word 'black' represents the event horizon, which is what makes the black hole completely dark. The word 'hole' represents the singularity, which makes part of spacetime cease to exist. But 'hole' also evokes the feeling of falling into a physical hole, in this case a hole that is impossible to get out of. When Schwarzschild wrote down his formula, however, the term 'black hole' did not exist and he just referred to his newly found equation as 'the gravitational field of a mass point' (we'll look into the origin of the name 'black hole' in more detail in Chapter 10).

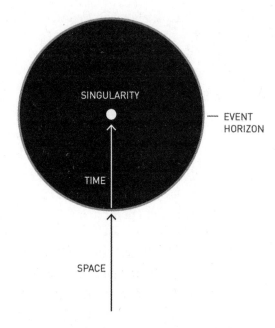

The structure of a black hole as described by Schwarzschild's formula. The event horizon forms the surface of the black hole. Inside the black hole, the singularity is a future point in time that all infalling matter will encounter.

The larger a black hole is, the more time it takes to reach the singularity after passing through the event horizon. In a black hole with a mass equivalent to that of the Sun, this would happen, according to Schwarzschild's metric, in just a couple of microseconds. If the mass is several million times larger – as with Sagittarius A* – it would take thirty seconds or so. For a black hole with a mass ten billion times that of the Sun, someone falling freely within the black hole would reach the singularity after nineteen hours. In this sense, black holes can be less dangerous than stars and planets. If you were falling towards the Sun, you'd burn up long before reaching its surface, but if you were to fall towards a gigantic black hole, you could both survive the passage into it and see what was happening inside.

As we saw in the Prologue, you would eventually be torn to pieces. Your body would be drawn out in one direction and squashed together in another. Stephen Hawking called this effect 'spaghettification', because everything that falls into a black hole gets elongated and pressed like a piece of spaghetti.[11] The point at which this spaghettification occurs depends on the size of the black hole. For a small black hole you might start getting stretched out before you've even passed the event horizon. But, for a very large one such as Sagittarius A*, all parts of your body would fall at the same rate towards the black hole. Therefore you could journey into it without initially sustaining any damage.

When you pass the event horizon you wouldn't notice anything in particular, just as nothing in particular happened in the metaphor with the river and the waterfall when the rate at which the river flowed grew stronger than the fastest human swimming speed. You'll go on travelling through the dark. In the end, however, you will begin to be drawn out and compressed inside it.

Because the singularity is in the future, you will not be able to study it. Furthermore, it is impossible to know what happens at the singularity using the Schwarzschild metric. In answer to the question 'What happens to the curvature of space and time there?' it spits out 'infinity'. That's the formula's way of saying, 'Something extreme happens here and I can't explain what it is.'

Just a few weeks after Einstein hit upon the equations that unite space, time and matter into a dynamic spacetime, Schwarzschild showed how they indicate a point where space, time and matter are torn apart. A black hole is the geometry of spacetime taken to its absolute extreme – to where it ceases to apply. Inside the darkness of the black hole, Einstein's theory of space and time breaks down. New quantum effects could arise that the general theory of relativity fails to describe. When Einstein's theory is used to describe the gravitational field of large objects such as stars and planets, theorists do not need to take into account quantum mechanics, which describes phenomena that occur on the minute scale of atoms. But close to the singularity, the gravitational field changes dramatically at very small scales, potentially making the quantum nature of space and time important. Physicists therefore need a quantum theory of gravity to explain the final stages inside a black hole. Exploring such a theory and thereby understanding the mystery of the singularity is one of the highest priorities for theoretical physicists. Despite decades of research into different quantum gravity theories, such as string theory or loop quantum gravity, and their application to black holes and singularities, no general consensus exists concerning the final fate of matter that falls into a black hole.

EINSTEIN'S REFUSAL

With Einstein's encouragement, Schwarzschild summarized his results in two scientific papers published in January 1916.[12] They analysed the possible gravitational field inside and outside a star, as described in Einstein's new theory. But Schwarzschild did not live to see that his newly discovered formula contained far stranger consequences, and that it would lead to the prediction of black holes existing in space. Welts and sores had begun to spread in his mouth and over his skin: he was suffering from the autoimmune disease pemphigus vulgaris. His immune system was attacking his own body.[13]

Schwarzschild went on making calculations in spite of the pain. He wanted to understand the implications of Einstein's theory for the conditions inside a star. He also threw himself into

BEYOND THE EVENT HORIZON

an analysis of the new quantum mechanical description of the atom. But his illness worsened and in the spring of 1916 he was admitted to hospital in Potsdam. He died on 11 May that same year.

Einstein saluted Schwarzschild's scientific contributions at his funeral.[14] But privately he expressed doubts. He wrote to a friend that Schwarzschild 'would have been a gem, had he been as decent as he was clever'.[15] As a pacifist, Einstein didn't like the fact that Schwarzschild had taken part in the war.

Nor did Einstein like the consequences of the Schwarzschild metric. The more scientists analysed the formula, the clearer it became that it described a very strange object indeed. It was questionable whether such an object could exist at all. Einstein didn't think so. He refused to accept that matter could be concentrated into such a small volume that a black hole would form.

About a month after the Second World War broke out, Einstein published a paper in which he wrote that the event horizon '[does] not exist in physical reality'.[16] However, other physicists were becoming increasingly convinced that black holes can arise upon the death of a star. Black holes became part of the answer to one of the oldest questions we have posed while gazing up at the stars: what happens to them when they stop shining?

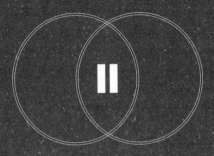

BLACK HOLES IN THE DEPTHS OF SPACE

They've circled each other for millions of years, slowly, at a distance, as though trying to catch the scent of their mutual attraction. As they approach each other, their pace grows incessantly. The two neutron stars have locked into a common course. Their compact circular march stamps vibrations through spacetime, gravitational waves that echo through the universe, presaging the crash that is to come.

At last they are so close that their neutron bodies deform. They swing round ever faster, nudging against each other and finally slamming into one another at close to the speed of light. In a violent cascade of energy their quark contents are torn apart and then crushed together. A radioactive witches' brew of star-dusted particles spews forth in a shower of gamma sparks and X-ray rain. The collision forms new elements that enrich the cosmos – but of the nova all that remains is night.

What's left is a black hole. Around its axis, space spins and twists several thousand times a second. The colliding neutron stars have given rise to a spacetime maelstrom.

Chapter 6

· · · · · · · · · · · · · ·

DYING STARS AND
SPACETIME VORTICES

As I'm in the middle of editing this book, a notification pops up in my inbox. 'Most massive stellar black hole in our galaxy found', the email subject line says. New records like this are constantly being set. In this case, a press release from the European Southern Observatory (ESO) issued on 16 April 2024 informs me that astronomers have found evidence of a black hole in the constellation Aquila.[1]

The space observatory Gaia identified it by observing a star 'wobbling'. This was a clear indication that the star had a dark partner invisible to the telescope. Follow-up observations by the ESO-operated Very Large Telescope in Chile confirmed that this partner weighed thirty-three solar masses (in astronomy it is customary to use the Sun as a unit of mass, as I will in this and subsequent chapters). They determined that it was, in all probability, a black hole. Once again, John Michell's more than two-hundred-year-old method for confirming the presence of dark celestial objects had produced results.

The black hole is called Gaia BH3. Aside from the hulking Sagittarius A*, it is, at the time of writing, the largest black hole astronomers have observed in the Milky Way. It's also one of the closest. At only 2,000 light years away, it is as close to Earth as some of the stars we can see in the night sky.

But it would be impossible to travel to Gaia BH3 to study it. The space probe Pioneer II, which NASA launched in 1973, will reach the star Lambda Aquilae in the constellation Aquila in four million years. That's how long it can take one of humanity's spacecraft to reach a nearby star, and the black hole is almost twenty times more distant. While we're on the subject of nearby black holes, the closest, at least of those we know about at present, is 1,560 light years away. This one is known as Gaia BH1. It weighs nine solar masses, and, as its name indicates, was also discovered using the Gaia observatory.

Astronomers estimate that there are more than a hundred million black holes like Gaia BH1 and BH3 in the Milky Way.[2] They all have a shared origin: they were once shining stars. Now we'll look at one of the ways a star's life can end, and at a new property black holes can obtain.

WHITE AND BLACK DWARFS

The Sun's rays caress my face. After a long, dark winter followed by a grey, rainy spring, I can finally sit outside to work. The seagulls and ducks on the shore of the lake I'm sitting by seem to have new energy too. They fly around, preen, chatter and seem, just like me, to have been missing the sunshine.

But the Sun won't shine forever. One day its energy reserves will run out. In the Sun's interior, atomic nuclei are whipping about at immense speeds. When they hit other nuclei they can enter into an atomic union and form a new nucleus. This process is called *fusion*. In the Sun 600 million tonnes of hydrogen are transformed into 596 million tonnes of helium every second. According to Einstein's famous formula $E = mc^2$, the difference of four million tonnes is transformed into energy that, among other

DYING STARS AND SPACETIME VORTICES

things, flows out of the Sun's interior in the form of an intensely bright light. It can take 100,000 years for this light to find its way through the Sun's gaseous mass to its surface. Once there, it rushes on through the cold void of space, reaching Earth after just eight minutes and, ultimately, me and the gulls and ducks on the lake.

But in around five billion years this internal fusion will cease as the internal supply of hydrogen is depleted. At that point, the Sun's core will consist of helium, which can, in turn, fuse to make carbon. After that the temperature will no longer be hot enough to continue the process of fusion into heavier elements. Stars that are more massive than the Sun have a higher internal temperature and, depending on their mass, can also transform carbon into neon, then oxygen, silicon and other elements up to iron. The fusion of iron into heavier elements consumes more energy than it releases, so it can't sustain itself under normal conditions. So, sooner or later, all the stars in the universe will reach the end of their fusion process.

Just as a bonfire leaves behind ash, the stars leave something behind when their fusion heart stops beating. Physicists and astronomers were kept busy for many years immediately before and after the Second World War trying to find out what this something might be. They wanted to find out what happens to a star when it reaches the end of its life.

To do this they needed to analyse two tendencies that govern the inner lives of stars. The situation is reminiscent of Faust's outburst in Johann Wolfgang von Goethe's play:

> Two souls, alas! are housed within my breast,
> And each will wrestle for the mastery there.[3]

In a mirroring of Faust's struggle, the two tendencies within a star – expansion and contraction – are pulling in opposite directions.

FACING INFINITY

The star is made to contract and shrink by the huge gravity of its matter. It is made to expand by the immense pressure created by the light rushing out of it and by the agitated movement of its gaseous mass. In the Sun, these two tendencies are balanced, but when fusion ceases, some of the star's internal pressure will be lost. The Sun will start to collapse – but into what?

We can find part of this answer if we look to the brightest star in the night sky, Sirius. Astronomers examining the star discovered that it had a stellar partner, known as Sirius B. This partner perplexed early-twentieth-century astronomers. It had a high temperature, but its total light output was meagre. This implied that Sirius B was much smaller than ordinary stars. When astronomers used Kepler's formula to calculate its mass, they were astonished. It weighed almost as much as the Sun, but its diameter was almost as small as Earth's. A teaspoon of its stellar material might weigh a whole tonne – it was like finding a mouse with the mass of an elephant. Such stars were given the name 'white dwarfs', white because their radiation was so hot, dwarf for their diminutive size.

Astronomers realized that a white dwarf was a star's final form. The Sun too will end its life in this way. First it will swell to become a red giant as the helium transforms into carbon in its core. Then its great plasma storms will envelope Mercury, Venus, and possibly even Earth. But when the internal pressure of fusion abates, it will begin to collapse, shrinking in size as its matter is squeezed tighter and tighter. In the end, the nuclei and electrons of its atoms will be so densely packed that a new kind of quantum mechanical effect will arise. A law of nature known as the Pauli exclusion principle states that no two electrons can occupy the same quantum mechanical state. Like a crowd forced together from all directions there is a limit to how tightly electrons can be packed. Their speed increases as they compete for

DYING STARS AND SPACETIME VORTICES

the little space that is left to them, and, just as a crowd grows more and more agitated as there is less space, the electrons start to push back. The quantum mechanical pressure exerted by the electrons acts against the inward pull of gravity. Expansion and contraction balance each other once again.

Most of the stars in the universe will end up as white dwarfs. They are like healthy pensioners, still vigorous in their old age. Even if the fusion within them has stopped, they go on shining because of all the excess heat they generated earlier in their stellar lives. As they beam out this energy, they gradually cool, and after more than ten trillion years they will be as cold as the space surrounding them. Scientists have speculated that they could then become black dwarfs. The carbon atoms inside them align into a kind of crystal, like an incredibly dense diamond. No stars have yet reached this stage, because it would take more time than the present age of the universe for a white dwarf to radiate all of its stored energy. And so black dwarfs remain a speculation, though it is exciting to imagine an incomprehensibly distant future where space could be filled with black, spherical crystals.

A white dwarf that can become a black dwarf is one of the answers to the question of what happens to a star when it nears the end of its life. But not all stars become white dwarfs. In the early thirties the nineteen-year-old Indian Subrahmanyan Chandrasekhar discovered that there is an upper limit to how massive a star can be and still become a white dwarf. If its mass is more than 1.4 solar masses, it will no longer be stable, because the gravitational force from its stellar matter will be stronger than the quantum mechanical pressure. For this discovery, Chandrasekhar was awarded the Nobel Prize in Physics in 1983. He had shown that if a star was sufficiently massive, it would have to become something else at the end of its life.

NEUTRON STARS

On 4 July 1054 a light flared in the sky in the region of the constellation Taurus.[4] Chinese and Japanese astronomers noted that this 'guest star', as it was called, could be seen in broad daylight. Initially the light was as bright as Venus and it was visible for almost two years before it vanished. In Egypt, the doctor Ibn Butlan claimed in alarm that the temporary star had led to low water levels in the Nile, the death of citizens in Cairo and Damascus, devastation in Iraq and the spread of disease.

Today we know that this point of light, which frightened and fascinated people both in the East and the West, was a supernova: an explosion that can occur at the end of a star's life. When fusion stops, a star of sufficient mass can implode in an instant. Vast quantities of energy are released as all its stellar material crashes into its centre. At the moment of the explosion the supernova can shine as brightly as the whole galaxy it resides in. A supernova is a star's grandiose way of saying farewell to its glittering life.

In 1921 the Swedish astronomer Knut Lundmark suggested that traces of the supernova noted by the Japanese and Chinese astronomers in 1054 should still be visible. The famous, beautiful Crab Nebula consists of a great cloud of dust that glows turquoise, yellow and blue.[5] Lundmark realized that this dust had come from the matter sent out by the exploding supernova.

In the depths of the Crab Nebula, astronomers can see what the supernova left behind: a neutron star. This is an extremely dense, almost perfect sphere with a diameter of less than twenty kilometres. A teaspoon of its neutron matter weighs as much as a whole mountain. The neutron star is created out of the inner core of the exploded star. When its mass is greater than the so-called Chandrasekhar limit of 1.4 solar masses, the quantum mechanical pressure of the electrons cannot prevent gravitational

DYING STARS AND SPACETIME VORTICES　107

implosion. The stellar matter is pressed together so densely that an atomic transformation occurs: electrons and protons are squeezed together so much that they are recast as neutrons. The imploding star becomes a kind of gigantic atomic nucleus made of neutrons. Just as a white dwarf is stable as a result of the quantum mechanical pressure of its electrons, a neutron star is stable because of the pressure from the neutrons.

In 1934, just two years after the British physicist James Chadwick had discovered the neutron,[6] the astronomers Fritz Zwicky and Walter Baade suggested the existence of these neutron stars.[7] They also thought, quite correctly, that a neutron star could be one of the possible end products of a supernova. But as with white dwarfs, neutron stars cannot be just any size. The limit is set at a little over two solar masses. A star more massive than this at the moment of collapse must continue to implode. The question is, into what?

You've probably guessed the answer already: a black hole. But let's pause and consider what is actually at stake. When a star implodes there is a struggle between gravity, which makes the stellar material contract, and the other forces of nature, which can prevent this collapse. The question is whether a star's gravity can become so strong that its matter starts to collapse without being able to re-establish a balanced state. If this were to happen, gravity would be given free rein and all the stellar material would be squashed together into a point. As we've seen, Einstein doubted this possibility. Arthur Eddington, one of the interwar period's foremost astronomers, put it in dramatic terms when he said, 'I think there should be a law of nature to prevent a star from behaving in this absurd way!'[8]

But on 1 September 1939, the same day Hitler's troops invaded Poland, the US physicists Robert Oppenheimer and Hartland Snyder published a paper that has been called 'the most daring and uncannily prophetic paper ever published in the field'.[9] The field in question was astrophysics, the study of the physical and chemical properties of objects and phenomena in space, such as galaxies, stars and black holes.

Today Oppenheimer is primarily known as the 'father of the atomic bomb', because he was the scientific lead in US efforts to build one during the Second World War. One reason the Pentagon chose the intellectually multifaceted physicist for this task was his groundbreaking research into the internal structure of the atom. Before the war broke out, Oppenheimer and two of his colleagues had applied this new knowledge to demonstrate that there must be a mass limit for neutron stars.[10] He and Snyder then used Einstein's equations to calculate what happens when a star goes into indefinite collapse.

They showed that, in just a few seconds, a star can shrink to the size of Earth, then that of a continent, a country, a city and, eventually, a point. At the same time, the imploding star's gravitational force becomes such that no light can leave it. Space and time shut themselves up inside one another. Nothing is left of the brilliant luminosity; only compact darkness remains. A black hole is born.

Oppenheimer and Snyder discovered a strange phenomenon that arises during this collapse: an external observer of the implosion will never see how it ends. It's as though the star's last moments are transformed into an eternal, unchanging present. The black hole's birth is in turn postponed until an infinitely distant future. Oppenheimer and Snyder discovered that *if* someone were to be floating nearby a star that was collapsing into a black hole, it would appear that the collapse was never fully achieved, even though for the stellar material itself the breakdown would take just a few seconds. This is the seemingly paradoxical consequence of a situation in which space and time are curved so powerfully that a black hole is produced.

DYING STARS AND SPACETIME VORTICES 109

Several physicists doubted Oppenheimer and Snyder's result. The two men's calculation was based on a number of simplifications. Most importantly, to be able to perform the difficult calculations they had assumed that the pressure within the star would be insufficient to resist the collapse, but this was precisely what needed to be proved.

In the end, determining whether a star could collapse and form a black hole was made possible by the development of one of humanity's most destructive creations: the atomic bomb, the very work that secured Oppenheimer a place in history. Many of the physicists who had worked on the US's atom-bomb programme known as the Manhattan Project, and its counterpart in the Soviet Union, subsequently turned their attention to imploding and exploding stars. After all, stars, supernovas and nuclear weapons had a good deal in common: high temperatures, densely packed matter, fusion and fission processes and, not least, explosions.

The physicist John Archibald Wheeler, a professor at Princeton University, was the one who ultimately answered the question of whether a star can become a black hole.[11] Today he is considered one of the foremost black hole scientists, but to begin with, he too doubted their existence.

During the Second World War, Wheeler was responsible for the production of the plutonium needed for the atom bomb. After the war he continued with his work for the military, tasked with undertaking the calculations for the world's first hydrogen bomb (which releases energy not through fission, where heavy atomic nuclei are split, but with the aid of fusion, where light atomic nuclei combine). To analyse the physics of the hydrogen bomb, Wheeler used the computer MANIAC (Mathematical Analyzer Numerical Integrator and Automatic Computer). It had been developed during the Second World War, weighed almost half a tonne, and was used to decode encrypted messages, calculate ballistic arcs and, later, to analyse nuclear weapon explosions.

Together with his colleagues B. Kent Harrison and Masami Wakano, Wheeler used MANIAC to figure out what happens to matter as its density continues to increase. *If* there is sufficient pressure inside the matter, no black hole can be created. But if the quantum mechanical forces acting between all the nuclei are not enough to hold out against gravitational collapse, it is unavoidable that a black hole will form.

MANIAC's computational capacity seems ludicrously small compared to today's computers. When it was set up to play chess it did so on a board with thirty-six squares instead of sixty-four. Every input required twenty minutes of calculations. It had a memory of five kilobytes, about the size of an email. Wheeler described in his autobiography how they tested the reliability of MANIAC's results by hitting it with a rubber mallet and then getting the computer to perform the same calculations again. If there was a discrepancy in the results, that meant something was wrong with the computer.

In spite of its limitations, MANIAC managed to answer the question of whether a star can become a black hole when it dies. After numerous, drawn-out calculations the answer was clear: a star with a large enough mass cannot resist gravitational collapse. Its internal pressure is too low, its gravity too high. Wheeler and his colleague had shown that a star really can implode and hide itself behind a wall of distorted spacetime. He compared the collapsing star with the Cheshire cat in *Alice in Wonderland*, writing that 'One leaves behind only its grin, the other, only its gravitational attraction.'[12]

Today we know that a star of eight solar masses or more can explode into a supernova, eject a large part of its mass and leave behind a neutron star or a black hole.[13] Which alternative comes

DYING STARS AND SPACETIME VORTICES

to fruition depends on complicated processes within the star towards the end of its life. Mapping them is at the forefront of today's research into the creation of black holes.

But there is one important aspect we still have to consider as we explore the transformation of a star into a black hole, and it reveals surprising new phenomena. Let's return to our neutron stars. They were first discovered by the Northern Irish astronomer Jocelyn Bell Burnell in 1967, when she was studying for a PhD at the University of Cambridge. With the help of a newly built radio observatory just outside Cambridge she observed a series of regular radio pulses. Initially she couldn't figure out where they were coming from, and jokingly referred to the signals as 'little green men'. Perhaps they were created by an extraterrestrial civilization? But when she discovered more radio pulses from other parts of the sky she realized they couldn't possibly come from aliens.

Instead, she worked out that the radio pulses must be coming from neutron stars.[14] Just as Earth's magnetic poles do not coincide exactly with the planet's rotational axis, the magnetic poles from which a neutron star's radio waves are emitted do not always align with the star's rotational axis. Thus, when a neutron star rotates, its radiation sweeps around in space like the beam of a cosmic lighthouse and its radio signal can regularly hit Earth (if pointing our way). This kind of neutron star is known as a *pulsar*.

Bell Burnell's doctoral supervisor Antony Hewish was awarded the Nobel Prize in Physics in 1974 for the discovery of pulsars, while she herself went without recognition. This has repeatedly been a cause of criticism: why was the male professor and supervisor awarded the prize and not the female PhD student who had made the discovery? Several of Bell Burnell's colleagues dubbed the Nobel Prize the 'No Bell', but Bell Burnell herself has never complained about the situation.[15] 'I felt it would demean the prizes if they were offered to graduate students,' she said in an interview.[16]

One of the pulsars Bell Burnell identified was the neutron star in the centre of the Crab Nebula. Since her initial discovery, more than three thousand pulsars have been located. The closest is 400 light years away, the furthest in another galaxy fifty million light years away. Pulsar signals are extremely regular. They can rotate several hundred times a second, which means the speed at their equator comes close to the speed of light. Imagine a compact, rapidly spinning ball of neutrons that has been squashed into a volume equivalent to a small city.

The rapid spin of a neutron star arises from one of the fundamental laws of physics, which states that if a rotating object shrinks in size, the speed at which it rotates will increase. Picture a figure skater doing a pirouette on the ice. When the skater pulls in their arms, they spin faster. The same goes for a star – if it shrinks, the speed of its rotation increases.

This in turn means that if a star forms a black hole, it too ought to rotate. The Schwarzschild metric, however, provides no information as to what happens with a black hole that is rotating. This does not mean the formula is incorrect; it still gives us important insights into the workings of the event horizon and the singularity, but it does not give a full picture of what a black hole is. To find out what happens to space and time around a rotating black hole, physicists needed to solve Einstein's complex equations. No one succeeded – until the New Zealander Roy Kerr attempted it in the early sixties.

FIFTY-FIVE CIGARETTES A DAY

'Let's see, Jonas.'

The eighty-six-year-old Kerr looks into the camera. After a few obligatory technical hitches we've finally managed to meet

DYING STARS AND SPACETIME VORTICES

on Zoom. It's the first time Kerr has used the program. I've sought him out because I want to hear him tell the story of how he discovered the formula for a rotating black hole. It's a story of success against the odds.

'You sent several questions in advance,' Kerr continues in his resonant voice. 'You know what, forget the questions, and I'll take it from the top.'

Roy Kerr was born in 1934 in the little town of Kurow in New Zealand.[17] He grew up in Christchurch, and had a great talent, but also a major problem. University education in New Zealand was old-fashioned. When Kerr studied at Canterbury College, Christchurch, most of the books in the university library were out of date. The university didn't have the money for new ones, so Kerr was forced to read old, obsolete theories. 'I got completely bored with the eighteenth- and nineteenth-century mathematics that was being taught,' he says.

Kerr quickly went through all the assignments and showed them to his lecturer, who said that since Kerr had nothing better to do he should do all the calculations again. 'The advantage of the flawed teaching was that I didn't have to listen to what people told me,' Kerr says. 'I thought for myself.'

In order to develop, Kerr needed to find a way to leave New Zealand. He could have applied for a PhD at the University of Cambridge, but his instructors neglected to tell him. Instead, he had to wait several years for the next opportunity to arise, in the meantime warding off boredom with billiards, golf and boxing. 'I was hopeless,' Kerr recalls, 'I couldn't even see the punches that were hitting me.' He got knocked out several times, and in the end one of his professors told him to stop. It simply wouldn't do for one of the college's foremost mathematical brains to keep getting knocked about in this way.

In 1955 Kerr was finally able to leave New Zealand to embark on a PhD at Cambridge. He hoped he would now find himself in more dynamic, intellectual surroundings, but he was once again disappointed. His supervisor was absent and he was forced to depend on himself for intellectual development. Kerr became fascinated by Einstein's general theory of relativity and began to study it in depth. After finishing his doctorate he was given a chance to continue his research in the US, first at Syracuse University in New York State, and then at the University of Texas at Austin.

Around this time, many physicists and mathematicians were trying to solve Einstein's equations to find a formula for rotating black holes. No one succeeded; the equations were just too hard. There was even a group of scientists who believed finding such a formula was impossible. Kerr wasn't convinced. When he discovered that they had made a mistake in their analysis, he decided to have a go himself. The years of absentee supervisors and old-fashioned teaching had given him the confidence to solve problems independently. 'I believed in myself,' Kerr says. 'I knew that I could do these calculations and that my method was better than others.'

Kerr spent several weeks wrestling with Einstein's complex equations. 'They're ghastly,' he says. Finally, though, he succeeded. He had found a formula that gave the correct solution, but he wasn't entirely sure it actually described a rotating black hole. Kerr's boss, the Austrian-US physicist Alfred Schild, was delighted. 'He knew that many people had been hunting for such a formula for decades,' Kerr recalls. 'He was more excited than I was.'

Schild encouraged Kerr to investigate the new formula's implications for the rotation of black holes. He and Kerr entered Kerr's office. Schild sat down and packed his pipe with tobacco.

DYING STARS AND SPACETIME VORTICES

Kerr lit a cigarette. 'I could smoke up to fifty-five cigarettes a day,' he says. While he smoked, he worked through his formula's predictions for the new spacetime object he had discovered. Finally, he turned to Schild and said, 'It rotates!'

Over the following weeks, Kerr, Schild and other physicists explored the mathematical consequences of Kerr's solution. Solving Einstein's equations mathematically and finding a new formula was one thing. The next step, understanding that formula's implications, might be just as hard. When Kerr and his colleagues analysed the new formula in detail they realized it held several surprises.

ROTATING SPACETIME

When moons, planets, stars and galaxies rotate, matter is spinning in space in some way or another. But when a black hole rotates, it is not matter that is spinning, but space itself. Everything that comes close to the black hole is affected by this rotation, just as objects are drawn into a tornado in the air or a whirlpool in the sea. But there are two important differences. In a tornado or a whirlpool, it is air or water that is rotating, but around a black hole it is not a physical substance *in* space that is rotating, but the actual space. It's also possible, in a tornado or a whirlpool, to travel against the rotation if you have a plane or boat with powerful enough engines. Admittedly, you could also use a spaceship to travel against the direction of rotation of a black hole, but only up to a point. Close to the black hole, the rotation is so fast that nothing, not even light, can travel counter to it. Regardless of how powerful your spaceship's engines were, it would be dragged along. The infernal vortex where this happens is called the *ergosphere*.

But all is not lost to those who enter the ergosphere. It is possible to navigate out of it and away from the black hole; as we saw before, it's only once you pass the event horizon that escape becomes impossible. 'It's best not to go in,' warns Kerr about the event horizon, 'because if you do, you won't come back.'

Inside the event horizon there's another surprise in store. 'There seems to be a second horizon,' says Kerr. 'All the rays of light that go through the outer horizon continue to the inner horizon.' Beyond this horizon spacetime becomes even more distorted. If a black hole does *not* rotate, the singularity is a point in the future. Kerr discovered that with a rotating black hole, the singularity goes from being a point in time to an extended form in space. 'If you go to the inside, there seems to be a ring singularity there,' he explains. The spacetime geometry around this singularity is peculiar. There are closed time-like curves within which future and past appear to meet. Take Kerr's formula to extremes and it looks as though paths to new universes open up within the inner horizon.

When I ask Kerr what this actually means, he replies, 'It means nothing at all!' The inner horizon is unstable and the formula provides no answers as to what happens to the rotating stellar matter inside the black hole. But he, along with most other physicists, is convinced that it correctly describes what it looks like from the outside.

As the black hole rotates, the rotation pulls the outer event horizon along for the ride, deforming it in the process. The black hole's surface becomes flattened at the poles, and the faster the rotation, the closer the inner horizon comes to the outer one. When the rotation reaches the speed of light, Kerr's formula warns that spacetime itself appears to turn inside out. The event horizon vanishes and the singularity within becomes visible to outside observers.

DYING STARS AND SPACETIME VORTICES 117

Most physicists believe that this kind of situation could never actually occur. The US physicist Kip Thorne has demonstrated that interactions and radiation from the matter moving around a black hole make a rotation faster than about 99.8 per cent of the speed of light unlikely.[18] Furthermore, Roger Penrose has formulated the *cosmic censorship hypothesis*, which states that the singularity is always surrounded by an event horizon. No one has yet succeeded in proving Penrose's hypothesis completely, however, because of the complexity of Einstein's equations.

The most notable aspect of Kerr's formula is that astrophysical black holes are characterized by only two numbers: their mass and the speed of their rotation.[19] (Electric charge is also a characteristic of black holes, but any black hole with an excess charge would quickly attract particles with the opposite charge and become electrically neutral. Therefore this number is usually left out when discussing black holes that actually exist.) Compare the characterization of the black hole above to, for example, a description of the human body. You would have to reel off a long series of traits, such as eye colour, hair type, wrinkles, nose size, bone structure, height, weight, scars, tattoos and so on. And these are only the visible traits. If you wanted to give a completely exhaustive description, you'd need to provide a breakdown of the state of the more than one trillion cells that make up the body. And each of those cells consists of over 10^{27} atoms (1 followed by 27 zeros). You would also have to describe the state of those parts of the body that do not consist of cells. Essentially, an exhaustive description of a human being's body is impossible. But for a black hole, the reverse is true: all you need do is provide those two values. In a way, the fact that this is enough to determine the properties of a black hole makes them the simplest objects in the universe.

A BUNCH OF HUNGRY LIONS

After Einstein published his field equations in the autumn of 1915, it took only a few weeks for Karl Schwarzschild to find the formula for non-rotating black holes. Then, because of the complex mathematics, almost half a century passed before Roy Kerr found the formula for rotating black holes. His discovery was a major achievement, but when he came to present his formula at a conference in Texas the week before Christmas 1963, he was disappointed. 'I arrive at the meeting and hear that Roger Penrose will give a talk about my formula,' the eighty-six-year-old recalls. 'Then I got pissed off. It was like I had thrown a huge piece of raw steak in front of a bunch of hungry lions, and everybody wanted a bite of it.'

Penrose had recently developed new mathematical techniques that demonstrated how black holes formed. He was known as a mathematical prodigy, and his results would eventually earn him a Nobel Prize in Physics. He was also an experienced and well-established speaker, which may have been the reason why the organizers chose him for the presentation.[20] But Kerr wanted to introduce his formula by himself. The organizers agreed to this, but only allotted him ten minutes. When he began to talk, several delegates left the room to take a break. Some went out for a smoke. Others had a nap in the conference hall. Few seemed to realize what a revolutionary formula Kerr was presenting.

'I was very annoyed that they just sat there and ignored what I was saying,' says Kerr. When he had finished, the Greek physicist Achilles Papapetrou got up and started berating the audience. 'He was furious,' Kerr recalls. 'He shook his fist and told them that we had been looking for a solution like this for forty years and that they should listen.'

DYING STARS AND SPACETIME VORTICES   **119**

The importance of Kerr's formula became clear just a year later. That was when the first signal from a black hole in the Milky Way showed up – and it was rotating extremely fast.

CYGNUS X-1

In 1964 scientists sent up a rocket from a military base in New Mexico.[21] The aim was to measure X-ray emissions in space. The rocket went up into the atmosphere, its instruments registered X-ray photons and then it crashed back down to Earth. When the scientists analysed the recorded data they realized there was a powerful source of X-rays coming from the constellation Cygnus, and so they named this source Cygnus X-1. Exactly what was causing the emissions was unclear at first, but as astronomers continued to study the object, they began to suspect it was a black hole.

Minor fluctuations in the X-ray emissions indicated that this invisible object was small, and further observations showed that Cygnus X-1 was not alone: the presumed black hole was orbiting a gigantic star at a rate of one revolution every five days or so. After studying the system, astronomers concluded that the mass of Cygnus X-1 was much larger than that of a neutron star. Indeed, according to the latest observations it is around twenty-one solar masses.

The X-rays are emitted by matter that has been heated to several million degrees as it orbits close to the black hole. The faster Cygnus X-1 spins, the closer the matter can get to the event horizon without falling through it. The closer the matter gets, the faster it moves and the stronger the X-rays it emits.

The radiation from Cygnus X-1 indicated that it had one of the most extreme rotations astronomers have studied, right on the limit of what is possible according to Kip Thorne's calculations.[22] This means Cygnus X-1 spins on its axis around eight hundred times a second. Floating in proximity to such a maelstrom in spacetime must be an incredible experience. A compact darkness, over 120 kilometres in diameter, that is spinning so fast that, were it able to produce sound it would be at a pitch equivalent to 800 hertz – a little like the lower notes we can reach when we whistle.

ONE LAST JOURNEY

'I hope I can go on NASA's first flight to Sagittarius A*,' Kerr says as we're rounding up our conversation. 'I figured the least they could do is take me there. Unfortunately, it's going to take a long time and I'm getting a bit old. I can do it, but I'm not going to go through the event horizon.' Kerr's joking of course. NASA can't send anyone to Sagittarius A*. It's hard enough for people to travel to the Moon – but isn't it a tempting thought? To see what goes on inside the Milky Way's central black hole with your own eyes?

Kerr demonstrated that Einstein's theory predicted rotating black holes. But Einstein's theory held the prospect of something else too, something he himself initially doubted: that space and time can vibrate in the form of gravitational waves. Cygnus X-1 and Gaia BH3 are not alone, they are each part of a double system with another star. But it's even possible for two black holes to circle one another. When this happens they send out gravitational waves that can be measured on Earth. Curious about how such observations are undertaken, I travelled to a desert in the USA to find out more.

Chapter 7

. . . .

A COSMIC
SYMPHONY

On a warm day in May 2023 I am driving along Route 4 S in Washington State, in the north-west of the USA. Reddish-brown desert sand and scrubby bushes spread out on either side of the road. The landscape has changed during my journey. I started in Seattle and have been driving south-east for many hours, passing forests of towering arborvitae and Douglas firs before vistas of snow-clad peaks inhabited by pumas and lynx took over. I ate lunch near the town of Snoqualmie where David Lynch filmed *Twin Peaks*. Then the snow and forest were replaced by dry plains and exposed hillsides. Here, in the middle of the Hanford desert, is where scientists pick up the signals from colliding black holes millions of light years away.

These signals are known as gravitational waves. They resemble ordinary waves, except in one key aspect. Waves in the air can be heard as sound. Waves in the sea can be swum in. We can see electromagnetic waves as light, and the Earth's waves can be felt when the ground shakes in an earthquake. Unlike all of these waves, which move through media that *exist* in space and time, gravitational waves are vibrations *in space and time themselves*.

Gravitational waves are created when masses accelerate, but it's hard to alter spacetime. Compare it to a still pool of water. If you were to throw a cork into it, only gentle ripples would appear on its surface, but replace that cork with a stone and you'll get larger waves. The heavier the stone, the greater the waves. The same is true of gravitational waves: the more massive and compact the accelerating object, the more powerful the gravitational waves produced. So the densest objects – such as neutron stars and black holes – will create the biggest waves.

LASER BEAMS AND TAMPER-PROOF STICKERS

I turn off Route 4 S and onto Hanford Route 10. After a few minutes I see my destination: a semi-circular tunnel. It's a couple of metres wide and four kilometres long. It looks as though a giant has dropped a huge cement toothpick in the middle of the desert. From the road I can see only one of the tunnels, but I know it's connected to a second placed perpendicularly to form an L. The tunnels are part of LIGO, the Laser Interferometer Gravitational-Wave Observatory.

LIGO measures gravitational waves by registering the vibrations of two mirrors. They hang at the far ends of the two long perpendicular tunnels, which have been pumped free of air to form one of the world's foremost vacuum systems. When a gravitational wave passes the facility it causes the distance between the mirrors to expand and contract. It sounds dramatic, but the effect is so small it's hard to imagine: the change in length is around a thousandth of the width of a single proton. This is the distance scientists have to measure to verify that a gravitational wave has passed through. It is equivalent to measuring a change of a hair's breadth in the distance from Earth to our solar system's closest star, Proxima Centauri!

The gravitational wave detector LIGO in the Hanford desert in Washington State, USA. Here scientists can observe the passing of gravitational waves created by merging black holes. A similar LIGO site exists in Livingston, Louisiana, and gravitational wave detectors have also been built near Pisa, Italy (named Virgo), and near Hida, Japan (named KAGRA).

Thanks to laser beams shooting back and forth along the tunnels it is possible to measure the changes in distance produced by the gravitational waves. The beams enter the two tunnels, bounce off the mirrors at each end and are reunited at the intersection of the tunnels. Just as other kinds of waves can enlarge each other or cancel each other out depending on whether their oscillations are aligned, the waves of a laser influence each other. When a gravitational wave goes past the complex, the paths along which the laser beams travel shift and the effect they have on each other changes. By analysing the variation in the strength of the beams where they intersect, scientists can determine whether a gravitational wave has passed through the desert in Hanford.

I park my car outside the entrance to LIGO, and am met by Michael Landry. He exudes a kind of calm authority, a personality trait that serves him well in his role as director of the Hanford facility. If he were an actor instead of a physicist, he would be perfect as a captain in *Star Trek*.

Landry leads me into the LIGO control room. 'There was an earthquake in New Zealand a few hours ago,' says a technician dressed in a blue shirt, shorts and trainers – an outfit as good as any when you're searching for black holes in distant galaxies. Earthquakes make the mirrors in LIGO's tunnels vibrate, disturbing the sensitive measurements. The walls in the control room are lined with computer screens showing diagrams, numbers, maps and graphs. One details whereabouts on Earth seismic activities have occurred. Another shows the size of the vibrations registered by LIGO. I can clearly see how the earthquakes in New Zealand have made the mirrors shake.

Aside from earthquakes, the mirrors are affected by everything from passing trucks and planes flying overhead to wind, lightning and the ocean waves hitting the coast. Even the gravity of the Moon can make the Earth's surface undulate by as much as ten centimetres.

In order to measure something as weak as a gravitational wave scientists must take all these disturbances into account. Just like the surface of a lake, the ground we stand on is never completely still, instead undergoing a constant succession of microscopic shifts. The mirrors must be as isolated as possible from these minute motions, and to achieve this, a delicate mechanical system is in place. As part of this system, the mirrors are suspended from fragile threads of quartz fibre so thin they would break if someone touched them.

'The mirrors have to be a factor of ten billion times more still than the ground at Hanford,' says Landry. 'LIGO is a really difficult experiment, and getting the mirrors that steady is one of the most difficult parts of it.'

But the experiment has been a success. On 14 September 2015 LIGO registered the first signal from two colliding black holes. 'We didn't know if colliding black holes existed,' Landry tells me, talking about the time before the detections. He was unsure whether they would be able to observe anything at all. But the signal Landry and his colleagues recorded bore witness to two black holes waltzing around one another in a distant galaxy 1.3 billion years ago. The two black spheres each had a diameter of around a hundred kilometres. When they circled each other they sent out gravitational waves taking energy along with them out of the binary system. The distance between the black holes decreased as they moved ever faster, circling ten times a second, twenty times, forty, sixty. In the end, they were a mere 300 kilometres apart, and their event horizons merged. Words cannot do justice to this strange encounter: they united, collided, became one. At almost half the speed of light, they crashed and created a new, larger black hole. Their union was so intense that the very core of spacetime quivered, sending out a last cascade of gravitational waves in every direction. For one single moment a burst of energy was ejected equivalent to ten times the amount released by all the stars of the universe – not in the form of light, but in gravitational waves.

The waves were flung away at 300,000 kilometres per second from this newly created darkness and onwards out of the galaxy. As they spread through the cosmos they grew weaker and weaker – until, after a little over a billion years, a small portion of them reached the Earth and the Hanford desert. LIGO's mirrors shuddered.

Before the scientists were able to announce they had registered the first ever signal from two colliding black holes, they needed to be certain that the vibrations had not been caused by something else. After excluding every possible Earth-bound source ('We even looked at lightning strike data from all over North America,' Landry says), their suspicions turned to another factor – the human one. The LIGO scientists had a protocol for secretly inserting an artificial signal into the system. Only a few people were able to conduct this kind of data injection. The purpose was to test that all the physicists analysed the data correctly, and that the observation method was satisfactory. Some of LIGO's scientists thought the signal was so strong it had to be the result of this kind of secret test. But as they proceeded with their analysis it became clear this was no artificial data injection.

That left just one possibility: they'd been hacked. 'Someone could have had an iPhone and injected waveforms into the electronics with clip leads,' says Landry. I picture a black-clad figure creeping across the desert at night and breaking into LIGO's tunnels. The shady visitor connects a cable to an electrical circuit with crocodile clips and sends off a signal that resembles two colliding black holes. 'No reasonable person would do something like that,' Landry points out, 'but we wanted to be sure that it hadn't happened.'

The hacker attack would also need to have happened in two places simultaneously. Aside from LIGO's Hanford facility, there is one in Louisiana. If a detectable gravitational wave passes the Earth, it will produce a result at one of the complexes a few milliseconds before the other. The scientists then use information about the order in which the signals were registered and the time lag to determine where in space the gravitational wave originated. If the scientists see the same signal in both locations, there is an increased possibility that it was produced by a gravitational wave and not an accidental fluctuation. That is, if it's not the work of hackers. 'We opened the electronics,' says Landry, 'inspected it, closed it and sealed it with tamper-proof stickers. Then we waited for the next detection. It came on Christmas Day.'

Landry had just sat down to Christmas dinner with his family when the phone rang. He was informed that LIGO had registered another strong signal. 'It was a big moment,' Landry tells me. The signals were not a hacker attack. They were not background noise. The gravitational wave chasers had recorded not one, but two strong signals from black holes. At a press conference on 11 February 2016 David Reitze, the director of LIGO, spoke the words many astronomers and physicists had been waiting for: 'Ladies and gentlemen. We have discovered gravitational waves. We did it.'

They had created a new way of measuring the universe. 'Gravitational Waves Detected, Confirming Einstein's Theory,' proclaimed the *New York Times*. Similar headlines were seen around the world.

In December 2017 three of LIGO's most prominent figures, Kip Thorne, Rainer Weiss and Barry C. Barish, a former director of LIGO, flew to Stockholm to receive the Nobel Prize in Physics. Michael Landry and several others from LIGO accompanied them. The first thing Landry did upon arriving in Stockholm was not to meet up with other physicists, however. Instead, after his wife had gone to bed, Landry went to an ice hockey training centre to meet the coach for the Swedish national team's goalkeepers. Landry is a goalie coach representing USA Hockey in Washington State and wanted to trade ideas. 'You have great teaching methods,' Landry says, looking at me. I nod politely and try to conceal the fact that, even though I come from Sweden, I know absolutely nothing about ice hockey.

PIANO KEYS AND GRAVITATIONAL WAVES

Before my trip to LIGO I chat to the ninety-year-old Rainer Weiss.[1] He talks rapidly and energetically to me over the internet from his home outside Boston.

Weiss's life story holds some fascinating lessons about what it takes to produce Nobel Prize-level physics. But his story is also one of music and its importance, both for our lives and for our understanding of the gravitational waves emitted by black holes.

Weiss was born in Germany in 1932. His father was a communist and a Jew, and when Weiss was six years old his family fled to New York to escape the Nazis. Growing up there, Weiss became enchanted by music. He listened to the New York Philharmonic on the radio every Saturday. He loved the symphonic sounds and set out to create his own audio system with a radio, record player and speakers. He got the speakers from a burnt-down cinema in Brooklyn. 'My friend and I had to climb ladders to get the movie theater's speakers,' Weiss recalls. 'The cones had burned out, but the magnets had survived.' Weiss replaced the cones and put the speakers in a cabinet. His parents' friends came round to have a listen: 'When they heard the sound, they wanted speakers like that.'

But, for Weiss, something marred the experience. 'Take the slow movement of Beethoven's *Appassionata*, all you hear is record noise!' In the thundering sections of the *Appassionata* the distortion was barely noticeable, but in the quiet second movement it detracted from his enjoyment. Weiss tried to build a system of electronic filters that could muffle the fuzz when the music entered the more muted passages, but he failed. He knew too little about mathematics and electronic systems, and so decided to study electrical engineering at Massachusetts Institute of Technology, also known as MIT.

A COSMIC SYMPHONY

His studies at MIT were a let-down. Everything was very strict, and Weiss became bored and dreamed instead of becoming a physicist. At the same time, a failing love affair with a woman in Chicago was draining all his energy. He was an unsuccessful student. 'I was a total disaster,' says Weiss, but notes how his passion for music was rekindled when his lover in Chicago sat down at the piano: 'She played the slow and sad Prelude in E-flat minor by Bach. It was heavenly to listen to.'

Weiss tried to learn the instrument himself, but he was already twenty-two years old. He regrets starting so late in life. 'I was unfortunately a victim of the Nazis,' he says, referring to the way his family were forced to flee Germany. 'Now I play every night, but I have no fluidity. I have no technique.'

A piano can teach us about the two most important properties of gravitational waves: their frequency and amplitude. A standard piano has eighty-eight keys. When you play a key, a wooden hammer strikes a metal string. The string starts to vibrate, forming waves of sound, which you hear as a note. The speed at which the string vibrates determines the pitch of the note, and the number of times the strings – and thus also the sound waves – vibrate per second is called the *frequency*, which is measured in hertz. The furthest-left key on a piano plays a note at 27.5 hertz, while the key at the right end plays a note at 4,186 hertz. The human ear can detect sound waves roughly in the range of 20 to 20,000 hertz.[2]

As well as having different pitches, a note can also be quiet or loud. If you hit a key hard, it produces a louder sound than if you hit it gently. The strength of the note is called the *amplitude*. Because we can't see sound waves in the air, it can be hard to imagine the amplitude of sound. But we can return to our image of a wave in a body of water. Let's imagine you're in the water. If the wave amplitude is low, it will create no more than a ripple;

if it is high, it will make your whole body bob up and down. Likewise, high amplitude sound waves will cause your eardrum to vibrate more strongly than low amplitude ones, meaning you hear a louder note. The pitch (frequency) and strength (amplitude) of a sound wave created by the piano are thus dependent on which key you play and how hard you play it.

All waves, including sound waves, electromagnetic waves and seismic waves, are characterized by their frequency and amplitude. The third key characteristic of a wave is its speed. Sound waves travel through the air at around 340 metres per second, whereas light waves travel through an electromagnetic field at around 300 million metres per second. This is why you see the lightning in a storm before you hear the rumble of the thunder.

Let's move on from the piano's sound waves to explore the gravitational waves emitted by a pair of black holes. According to Einstein's theory, light and gravitational waves move at the same speed. When two black holes rotate around one another they create gravitational waves at a frequency that is twice the number of rotations completed by the black holes every second. If they circle each other ten times a second, they emit a gravitational wave of 20 hertz.

The distance between the two black holes decreases as the gravitational waves carry energy away from them. Remember the figure skater spinning faster and faster as they pulled in their arms? The same thing happens here: the black holes spin faster around each other. The frequency and amplitude of the gravitational waves increases, leading to even more energy being removed from the system. This process is self-perpetuating: the stronger the gravitational waves emitted, the closer the black holes get, and the closer they get, the higher the frequency and amplitude of the gravitational waves they emit. This means that when two massive objects such as black holes – or neutron stars – circle each

A COSMIC SYMPHONY **131**

other and collide, they produce a characteristic signal, known as a chirp. Initially there is a low-frequency wave that can last for several million years while the two dense objects approach each other. In the final phase, however, the signal takes on a higher and higher frequency until the two objects merge in less than a second, creating a new, larger black hole. Then there is silence.

The exact frequency and amplitude of a given sequence of gravitational waves depends on the mass of the black holes or neutron stars. The smaller they are, the closer and faster they can circle each other. This means that small black holes emit higher-frequency waves than large ones. LIGO can measure frequencies from 10 to 10,000 hertz – notably close to the frequencies our ears can hear, though in the case of our ears, of course, it is sound waves and not gravitational waves that are picked up. If we were to float somewhere close to two black holes that were sending out gravitational waves at these frequencies, our bodies would be stretched and compressed at the same frequency. Since this would also affect our eardrums, we might be able to hear the waves passing.

In 1967 Weiss hit upon a way of observing gravitational waves. After failing in his studies, he got a job as a carpenter for a research group at MIT. He was given increasingly tricky tasks and after a while started conducting his own experiments, before embarking on a PhD and ending up as a research group leader. His technological expertise was all self-taught. When I ask him where his natural feel for electronics comes from, he says he doesn't know. 'It's completely intuitive,' Weiss says. 'My father or mother were not that way, and not my children either. I'm an anomaly.'

The head of Weiss's department at MIT wanted him to teach general relativity. 'It was terrible,' he recalls, 'because I didn't know general relativity. I had never done the calculations, but I couldn't tell that to the head of the department.'

Weiss was given three months to prepare the course. 'The students realized I was a total novice,' he says, 'but they were fine with it. They were very gentle.' Weiss taught and the students asked questions. They'd heard that the physicist Joe Weber at Maryland University claimed to have observed gravitational waves. Weber had tried to measure the waves using a two-metre-long aluminium cylinder. If a gravitational wave passed, the cylinder would begin to vibrate. The vibrations were extremely weak, but Weber insisted he had managed to measure them using electronic sensors. Other scientists were sceptical. Several research groups carried out similar experiments without being able to measure any gravitational waves. 'Weber was not critical enough of what he was doing,' says Weiss. 'He didn't scrutinize things that affirmed his result, while at the same time he scrutinized things that disagreed very hard.' Today the scientific consensus is that Weber never measured any gravitational waves, though he is still viewed as a pioneer, since he made more scientists realize that the waves were not merely a scientific concept but something that could be investigated through experiments.

Weiss's students were curious about gravitational waves and had asked how they could be observed, so one Sunday morning Weiss sat down at his desk, took out pen and paper and started sketching and calculating. He imagined two mirrors floating freely in space. A laser beam bounces between them, with the distance between the mirrors determining how long the laser beams take to travel back and forth. A passing gravitational wave would cause the distance to change, thereby altering the beams' travel time. If a scientist could measure this interval, it should be possible to confirm that a gravitational wave had passed the mirrors.

In 1972 Weiss published an internal report in which he laid out the fundamental principles behind his gravitational wave detector. One of the most important aspects of the MIT report was a description of all the potential disturbances – all the noise! – on Earth that could prevent these weak gravitational waves being detected. He wanted to avoid noise in the gravitational wave detector just as he had in his audio system. Weiss mapped out every possible effect: from ground vibrations to geomagnetic storms and cosmic radiation. He also detailed how the laser beams, the mirrors and the vacuum through which the laser beams were to travel might distort the measurements. Weiss's report was a milestone in the search for gravitational waves.

When the report was published, no one knew for sure whether gravitational waves or black holes existed. Weiss started collaborating with the physicist Kip Thorne. Thorne led a theoretical research group at California Institute of Technology (Caltech), had long hair, colourful clothes, and could easily have blended in with a group of California hippies. He and his colleagues were investigating which objects in the universe could create gravitational waves and what the signals would look like. He realized that it should be possible for a facility on Earth to measure gravitational waves from black holes, but at that point no one knew how often two black holes circled each other and eventually merged. Perhaps it was so unusual they would never have a chance to detect it, or perhaps it happened so often their facility would be able to observe several such events every year.

It was like the scientists were about to enter an uncharted forest, tasked with finding out how many birds lived there using only their song. Before they went in they would have no idea how many birds there were or how much tweeting they could expect to hear. Perhaps the birds would sing constantly, perhaps the whole place would be completely silent. In a similar way Thorne, Weiss and their colleagues had no idea how much they would 'hear' the black holes' gravitational waves. Was the universe completely silent? Or did the 'noise' of the colliding black holes ring out loud and clear?

An important step in the right direction came in 1974 when the astronomers Russell A. Hulse and Joseph H. Taylor Jr reported seeing two neutron stars moving both around and towards each other. The rate at which the neutron stars spiralled inwards exactly matched the prediction from the general theory of relativity in terms of how gravitational waves would carry energy away from the system. Hulse and Taylor's measurements were an indirect proof of the existence of gravitational waves. They were later awarded the Nobel Prize in Physics for their discovery.

In 1979 the National Science Federation, a US federal body that funds science and technology research, agreed to support LIGO. No one knew if the experiment would work. It required technology that didn't even exist when work on the project started. Several researchers protested: why not put the money into something with a guaranteed result?

In my work on this book I have interviewed Nobel laureates in Physics and scientists who've been touted as possible Nobel Prize winners. In every case a clear pattern has been discernible: they must keep their projects going not just for years, but for decades. The technology needed is not always available in the early phases, which means taking risks. Will technology develop quickly enough for the project to succeed? This difficult, risky work conducted over long periods comes with a destructive side: conflict. Everyone I have interviewed has spoken of upsetting, hard-to-overcome arguments over prestige, money and influence. The challenges wear away at the scientists' mental health. No surprise, then, that a 1989 report into the construction of LIGO begins with a quote from Machiavelli: 'There is nothing more difficult to take in hand, more perilous to conduct, or more

A COSMIC SYMPHONY

uncertain in its success, than to take the lead in the introduction of a new order of things.'[3]

Weiss, Thorne and hundreds of other scientists worked hard for decades to bring the LIGO project to fruition. And they succeeded: on 10 December 2017 Rainer Weiss, Kip Thorne and Barry C. Barish accepted the Nobel Prize in Stockholm, after proving that it was possible to observe gravitational waves from black holes. But what have the gravitational wave hunters discovered since then?

COSMIC FIREWORKS

Back to the LIGO control room. A technician hands me and Landry a glass of apple juice each. They are celebrating the reinitiation of their hunt for gravitational waves after several years of dormancy during the COVID-19 pandemic. They'll be joined in the hunt by the European gravitational wave detector Virgo, located near Pisa in Italy, and the Japanese detector KAGRA, forming a global network searching for signals from black holes.

'We're very excited,' says Landry, putting his glass of juice down on the table. Up until the pandemic, LIGO had detected over ninety gravitational waves. 'Now we're going to observe several hundreds of events over the next eighteen months. This will allow us to say things about the whole population of black holes orbiting each other and understand where they come from.'

Did the colliding black holes originate from two stars that were born together and then exploded in supernovas? Or did the black holes form separately in different parts of the galaxy, only to run into each other later? These are the questions scientists hope to answer with help from data gathered by LIGO, Virgo and KAGRA.

FACING INFINITY

Another question scientists want to get to grips with is why the black holes they have observed are so massive. The first gravitational wave they detected had been created by two black holes of around thirty solar masses each.[4] 'I was totally surprised,' says Landry. ('Where the hell did they come from?' was Weiss's reaction.) Up until that point, all the black holes observed by astronomers (aside from supermassive ones like Sagittarius A*) had been smaller and lighter than this. But the more gravitational waves LIGO and Virgo observed, the more they kept discovering these large, heavy black holes. 'We've seen black holes of almost a hundred solar masses and merger objects with almost 150 solar masses,' Landry tells me.

The work they are doing is thereby providing an insight into the kind of stars that can collapse to form black holes. Some scientists have even advanced the theory that black holes don't come from stars, but that they were created in the first moments of the universe (more on this in Chapter 9). Whatever the answer is, future observations will lead to deeper insights into the origin of black holes. 'This clearly underscores why we are looking for the gravitational waves: to find things you can't find in any other way,' Landry emphasizes.

On 17 August 2017 astronomical history was made. LIGO and Virgo observed a gravitational wave triggered by two neutron stars colliding in a veritable cosmic fireworks display. A second or so after the observation, two space telescopes registered a burst of gamma rays. Astronomers around the globe immediately began searching for the origin of the signals. With the aid of information from LIGO and Virgo they were quickly able to pinpoint the collision in the galaxy NGC 4993, 130 million light years away. Over seventy telescopes studied the afterglow of the collision. 'The event was a complete gift from nature,' says Landry. 'We were really lucky. We were ready, but we were also lucky.'

A COSMIC SYMPHONY

For the first time, astronomers could observe both the gravitational waves and the light from the same event. Because the electromagnetic signal and the gravitational waves arrived on Earth almost simultaneously, it provided strong evidence that gravitational waves and light travel at the same speed (as predicted by Einstein's general theory of relativity). Almost a third of all the astronomers in the world were listed as authors of the paper that described the event.[5] The gravitational wave signal was observed for a little over a hundred seconds and went from 20 hertz up to more than 500 hertz before vanishing into the detectors' background noise.[6] The two neutron stars may have been revolving several hundred times a second before they collided. In all likelihood, they became a black hole after the merger.

Astronomers were able to confirm that the explosion, called a kilonova, led to the formation of many elements heavier than iron. At a press conference, LIGO scientists explained that as many as ten Earths' worth of gold and platinum could have been created in the collision. This probably means that the gold in our jewellery came from colliding neutron stars, and that most of the heavy elements on Earth may have been created in such cosmic mergers.

But what is the long-term goal of this hunt for gravitational waves? Sheila Dwyer, an MIT scientist, sits down at our table in the control room. She starts telling me about her work on upgrading LIGO's detection capabilities. I nod and try to keep up with her explanation of something called squeezed light, a novel technique devised to reduce the noise from quantum vacuum fluctuations in the detectors. Despite my PhD in physics, I struggle to understand the more complicated aspects of LIGO's technical systems.

Suddenly Dwyer tells me something that really astonishes me. 'If the detectors are ten times more sensitive,' she says, 'we can observe all the stellar-mass black hole mergers in the history of the universe.' Seeing almost all the colliding black holes – all the way back to the real dark ages when there were no stars lighting up the universe. I can hardly believe it's true, but it is. 'It's kind of mind-blowing,' Dwyer agrees.

Since LIGO observed its first gravitational waves in 2015, scientists have seen an event roughly every three months, though this increased to one a month after an upgrade. The new observations that were about to start when I visited LIGO have led to the detection of two or three events every week. Planned upgrades will lead to the observation of gravitational waves every day.

There are plans to construct two new detectors in the USA and Europe – Cosmic Explorer (USA) and Einstein Telescope (Europe) – that will have tunnels around ten times the length of those at LIGO.[7] These observatories will see even more events, and their instruments will not have to wait to observe the signals: they will constantly be able to monitor the vibrations from black holes colliding in distant galaxies.

The universe is full of gravitational waves playing out in a cosmic symphony, but the lowest bass notes in this symphony are made by supermassive black holes. Their frequencies are so low it would take a detector bigger than Earth to measure them. To get around this, scientists intend to send the Laser Interferometer Space Antenna (LISA) up into space in the next decade.[8] LISA consists of three measuring stations that will be spaced at intervals of several million kilometres. Laser beams will zip back and forth between the stations. This 'LIGO in space' will be able to detect single gravitational waves from merging supermassive black holes, providing brand-new insights into the spacetime vibrations that fill the cosmos.

AMONG RABBITS AND RAVENS

I leave the control room to go and take a look at one of LIGO's tunnels. The sky is bright blue, and on the horizon the desert mountains lie in silent repose. After a short walk I come to a tunnel. Laser beams are ping-ponging back and forth inside, and outside there's a rabbit hopping about among the bushes. These tiny movements are hardly enough to disturb the sensitive instruments, but when LIGO first started its observations there was one signal that kept appearing. The LIGO scientists could not understand where it was coming from, but after much searching they were able to identify the cause: ravens. These smart birds had figured out that a cooling system beside one of the tunnels was leaking ice. The ravens pecked at the ice to get water, a precious commodity in the desert. A PhD student tried tapping the ice with a hammer. Sure enough, the same signal showed up, and the mystery was solved.

The Hanford desert is a place where you can find rabbits and ravens, snakes, porcupines, scorpions, prairie dogs and the traces of black holes colliding billions of kilometres away. My body, the rabbit in the bushes, the mountains on the horizon and our entire planet are continuously being permeated by the gravitational waves from these black holes. Most pass undetected, but some of them are caught by LIGO. Like the Polynesian seafarers who studied the patterns of the waves to navigate the oceans, astronomers have learned to use the shape of the gravitational waves to understand our universe.

My visit to LIGO is over. I bid farewell to Michael Landry, Sheila Dwyer and a few others, and travel away from their complex along Route 4 S. Country music twangs from my radio. As I drive through the reddish-brown desert landscape I think of a teenage Rainer Weiss tinkering with radios so he could listen to the music he loved. Thanks to the initiative he took with LIGO we can now listen to the sounds of space in the form of the gravitational waves created by black holes. So far we've heard only a fraction of this cosmic symphony. In the next decade, we'll have new instruments on Earth and in space that will help us hear more of this music. Perhaps that's why the ninety-year-old Rainer Weiss responded as he did when I told him I played the piano as a kid, but that I don't play much now.

'Play more,' he said. 'As you get older you find out how important it is.'

Chapter 8

. . . .

THE SHADOW
HUNTERS

One July evening in 1978, in Meudon just south of Paris, the astrophysicist Jean-Pierre Luminet was sitting at his desk.[1] He took a pen and drew a point on the paper in front of him. Then he drew another point, and another. Luminet tended to use drawing as a way of winding down after a long day at the Meudon Observatory, but this summer evening his drawing was not a flight of fancy but a scientific illustration. He was going to show what the region around a black hole looked like.

Several of Luminet's colleagues doubted the existence of black holes, but he was keen to study them more closely. He was skilled in the mathematical modelling of cosmic phenomena, and with the aid of a big IBM computer in the basement of the observatory, he wanted to model how a black hole would appear to someone floating just outside it. Although the black hole would be completely dark, gas whirling around it would emit light, rendering the darkness visible. Some of the rays of light would pass the event horizon and disappear forever; others would be affected by the black hole's gravity but still manage to escape and reach a distant observer.

On a thick stack of punch cards Luminet had written a program to calculate the distorted paths of the light. He had given the cards to a data operative who contacted Luminet a week later with the words he'd been hoping for: 'We have a result.' On a computer printout Luminet could see a kind of contour map outlining how brightly the gas would shine around the black hole. But he had to add by hand all the points of light the computer wasn't capable of printing out. The next two evenings were spent at his desk, as he filled in over ten thousand dots.

'Illuminer' means 'to illuminate' in French, and Luminet's name really did suit him: with his picture he showed how a black hole would be rendered visible by the glowing matter around it. Luminet pictured a circular area of darkness enclosed by a glowing disc. The left-hand part of the disc is moving towards the observer, the right-hand part is moving away, making the left side brighter than the right due to a phenomenon known as the *Doppler effect*. An everyday example of the Doppler effect is as follows: imagine you're standing by a road and an ambulance is coming towards you. You hear the pitch of its siren rise as it approaches, but the moment it passes you and starts travelling away, the pitch drops. Similarly, the light from the gas is brighter when it is travelling towards the observer than when travelling away.

Jean-Pierre Luminet's scientific drawing from 1978 of a hot gas swirling around a black hole. The illustration shows how the light from the gas is bent around the black hole. The illustration is based on computer calculations and Schwarzschild's black hole formula.

THE SHADOW HUNTERS

In Luminet's virtual image the observer is viewing the disc slightly from above. Because the light's path is distorted it looks like the disc runs around the upper part of the dark area. This is an optical illusion caused by the gravity of the black hole. It can bend the course of the light so much it is even possible to see what is behind the black hole. Within the darkness enveloped by the disc there is a faint ring, the so-called photon ring, created by light travelling along a last-gasp orbit just outside the black hole.

Luminet was enchanted by the extraordinary aesthetic of this illustration. He thought it looked like a black eye surrounded by a peculiar halo. Just after finishing the image he came across a poem by the French poet Gérard de Nerval called 'Le Christ aux oliviers'.[2] He was astonished to find that two verses from the poem, which was written in 1854, described, in almost prophetic terms, the image he had drawn:

> In seeking the eye of God, I saw nought but an orbit
> Vast, black, and bottomless, from which the night which there
> lives
> Shines on the world and continually thickens
> A strange rainbow surrounds this somber well,
> Threshold of the ancient chaos whose offspring is shadow,
> A spiral engulfing Worlds and Days!

A night that 'shines on the world' is undeniably a fitting phrase to describe the darkness of a black hole.

In a collaborative paper with his former supervisor Brandon Carter, Luminet published his image in the French journal *La Recherche* in 1978.[3] Around the same time, a group of US astronomers investigating the galaxy M87 had discovered a probable gigantic black hole at its centre, deemed to have a mass equal to several billion Suns. Luminet claimed that because of its size, this black hole offered the best opportunity to observe the kind of darkness he had depicted.[4]

When I interview Luminet, I ask what he'd thought the likelihood was of ever seeing a real-life version of the drawing he'd created. 'I was reading science fiction books at the time,' Luminet replies, 'so I thought that in several centuries we might have a spaceship that could pass close to a black hole and take a picture like that.'

But it didn't take a spaceship to get a real picture, nor did it take several centuries. In 2019 scientists working for the organization Event Horizon Telescope (EHT) said they had achieved something most people thought impossible. They had managed to photograph a black hole.

THE SHADOW OF A BLACK HOLE

The morning Sun blazes down on Granada, in southern Spain. I start to sweat as soon as I leave my air-conditioned apartment. I walk along winding cobbled streets, past café terraces and Moorish-style buildings. I follow the orange-tree-shaded bank of the Genil river towards my destination: the Parque de las Ciencias.[5] Hundreds of astronomers (such as Geoff Bower, who appeared in Chapter 2), physicists, engineers and data scientists have come together to spend a few boiling summer days in this giant conference centre. They are members of the projects Event Horizon Telescope and Next Generation Event Horizon Telescope.

'One of the main goals of the Event Horizon Telescope is to understand the properties of space and time around a black hole,' Sara Issaoun tells me. She is one of the world's

THE SHADOW HUNTERS 145

foremost – and youngest – experts in the depiction of black holes. I meet her in a break between presentations to find out how one becomes a black hole photographer and what Issaoun and her colleagues have learned about these objects. Sara Issaoun was born in 1994, close to the Algerian city of Tizi Ouzou. As civil unrest and safety risks grew in the country, her family decided to emigrate. At the age of seven, Issaoun swapped the heat of Tizi Ouzou for the chill of Montreal in Canada.

Her interest in astronomy was sparked by a school assignment to build a model of the solar system. She started wondering why the small planets were closer to the Sun, while the big ones were further away. She borrowed some children's books on astronomy from the local library and read about Mercury and Jupiter, distant stars, and the black holes that fill the universe.[6] Issaoun tells me that, from that point on, she knew what she wanted to do with her life: 'I wanted to be an astronomer.'

After leaving school she started studying physics at McGill University in Montreal. One day Andrea Ghez came to the university and described how she had followed the movements of the stars around Sagittarius A*. 'I saw how enthusiastic she was and how much she loved what she was doing,' says Issaoun. 'As a woman, it's important to see female astronomers who have a strong passion for their work. Ever since then, I've been interested in Sagittarius A*.'

In 2014 Sara Issaoun saw her chance to try life as an astronomer. Her parents were working in the Netherlands and Issaoun was going to stay with them for the summer. She contacted several Dutch professors and asked if she could do a summer project about space. Heino Falcke, a professor of astroparticle physics and radio astronomy at Radboud University Nijmegen, was quick to reply. He asked her to visit the university to discuss a potential project.

In his office, Falcke told Issaoun that he had one major scientific goal: to capture an image of a black hole. It sounds impossible, since the definition of a black hole is that no light can leave it, but together with the astrophysicists Fulvio Melia and Eric Agol, Falcke had shown that it would be possible to use radio telescopes on Earth to depict what they called the 'shadow' of a black hole.[7]

It all comes down to the interplay of light and gravity. When Einstein developed his general theory of relativity, he calculated how the curvature of space and time around a star or planet could alter the paths along which light travelled.[8] The gravity of an object in space would cause passing light rays to bend away from their rectilinear course. The English astronomer Arthur Eddington and his colleagues tested Einstein's prediction in 1919. During a solar eclipse it was possible to see the stars close to the edge of the Sun. Using photographs taken during an eclipse seen both from the island of Príncipe off the west coast of Central Africa, and the outskirts of the city Sobral in north-eastern Brazil, they were able to confirm that the Sun's gravity had bent the stars' light in just the way Einstein had predicted.[9] It was the first experimental proof of Einstein's general theory of relativity, and it made Einstein world-famous overnight.

While Eddington had shown that the Sun affected the light of the stars, Falcke and his colleagues wanted to show how a black hole affects the light from matter travelling around it. The powerful gravity of a black hole means that the ways light can travel around it are quite different to the ways light passes around the Sun. Hot gas circling a black hole sends out light in all directions, and it is possible for this light to travel once or many times around the black hole. Some of the light might leave the immediate surroundings of the black hole, while some of it will disappear behind the event horizon. To the observer, there will therefore appear to be an area of reduced brightness in the centre of the black hole. Falcke and his colleagues named this phenomenon the black hole's 'shadow'. Depending on the speed of the black hole's rotation, the diameter of this shadow can range from around 2.6 to five times the diameter of the event horizon.

Unlike Jean-Pierre Luminet's hand-drawn image, which showed what someone floating close to a black hole would see, Falcke and his colleagues had shown what the black hole's shadow would look like to a telescope on Earth. Seeing the shadow is like sneaking a look into the black hole's secret workshop, examining how matter is simultaneously pulled in and shot out, how plasma whirls around at tremendous speeds, how magnetic fields are twisted, and space and time curved. This makes it possible both to investigate the complex physics taking place around a black hole, and to ascertain whether the predictions in Einstein's general theory of relativity really hold true.

There were two black holes whose shadows Falcke and his colleagues particularly wanted to observe. One was Sagittarius A*, and the other was M87*, the black hole at the centre of the galaxy M87. The French astronomer and comet chaser Charles Messier gave the galaxy its name in 1781. He saw a dim light source in the night sky and noted it down as number eighty-seven on his list of celestial objects, without actually knowing what it was. Today we know that M87 – that is, the eighty-seventh object on Messier's list – is the largest member in a group of galaxies called the Virgo cluster. The galaxy is fifty-five million light years away, which means M87* is 2,000 times further away from us than Sagittarius A*. Despite this, the team believed it would be possible to see its shadow, as M87* is much bigger than Sagittarius A*.

There was, however, an almost insurmountable barrier: the observations would require an absolutely enormous telescope. The black holes are wreathed in clouds of glowing particles that block most of the light and make it impossible to see the shadow.

Only short radio waves, with a wavelength of around a millimetre, can manage the arduous journey through the clouds, interstellar space and finally Earth's atmosphere without faltering along the way. But calculations based on the laws of optics indicated that if a telescope capable of observing such wavelengths were to be able to resolve the shadows, it would have to be almost as big as Earth itself. The project seemed doomed to failure before it had even begun.

Then a solution presented itself: by combining the observations of several observatories across different continents, it would be possible to gather the same information as a single Earth-sized telescope might.[10] The US astrophysicist Sheperd Doeleman from the Center for Astrophysics | Harvard & Smithsonian had invested a significant part of his career in getting just this kind of project off the ground. He sometimes described it as taking a telescope the size of Earth, smashing it with a hammer, distributing its parts around the globe, and then joining together the observations of the various parts in a supercomputer. The technique is called *very-long-baseline interferometry* (VLBI).

When Doeleman was a PhD student at MIT in the late eighties and early nineties, several of his colleagues were convinced that there were supermassive black holes at the centre of most galaxies, even though no one had ever seen them.[11] Doeleman was driven by a desire to depict what was happening right up close to the event horizons of these giants. Only then would it be possible to confirm that the compact objects astronomers were observing were really black holes. So he visited observatories, tested new technologies, courted funders and gathered a growing band of scientists with the aim of pushing radio telescope technology to its extreme limits. No one knew if they would succeed. Perhaps the black holes were surrounded by gas so dense it couldn't be seen through at all. Perhaps they wouldn't be able to develop

THE SHADOW HUNTERS

technology at the level needed to observe the shadows. Perhaps clouds of space dust would make the images so blurry the project would be impossible to complete. Some measurements of the mass of M87* also indicated that the shadow would be too small to see.

In 2008 and 2012 Shep Doeleman and his colleagues had succeeded in observing extraordinarily small features in Sagittarius A* and M87* that suggested it might be possible to see their event horizons.[12] This gave Doeleman, Falcke and close to a hundred other scientists the confidence to found the Event Horizon Telescope project.[13] Doeleman was named the organization's founding director. The telescopes they wanted to use were in far-flung places, such as the Atacama desert in Chile, the icy South Pole, and mountain peaks in Mexico, Arizona, Spain and Hawai'i. Their success depended on them conducting exhaustive diplomatic work on several continents to secure observation time at the telescopes. They had to install advanced equipment, collect vast amounts of data and develop new image algorithms.

In spite of these challenges, Heino Falcke asked Sara Issaoun if she was interested in learning the techniques they were developing. She said yes. Three years later it was time to try taking the first picture of a black hole.

THE OBSERVATIONS BEGIN

At the end of March 2017 Issaoun was in a car with her colleagues, travelling towards Mount Graham, the highest mountain in Arizona. She was twenty-two years old, still a student, and one of the youngest members of the EHT. Over the last few years she had been learning more about astrophysics and VLBI, as the

EHT members had been campaigning for new equipment to be installed at various telescopes around the world.[14] Now she was on her way to the Submillimeter Telescope, a radio telescope operated by the University of Arizona.

After a journey of several hours through the desert, Issaoun and her colleagues took the turnoff marked 'Mount Graham'. As they approached the peak she caught glimpses of the telescope's antenna between the pine trees. It wasn't the first time she'd visited. Heino Falcke had taken her there a year earlier. 'I showed her how the telescope works,' he tells me when I speak to him in his office at Radboud University, 'and after a day I was out of a job. She literally took control of it. She loved the telescope, and it loved her.' When I ask Issaoun how it felt the first time she saw the Submillimeter Telescope's antenna, she replies, 'I thought, "Ten metres in diameter is not that big," until I stood in front of the telescope and realized how enormous it is.'

In order to capture images of black holes, the Submillimeter Telescope and the other seven telescopes spread across four continents would have to work flawlessly. The astronomers would also have to be lucky with the weather. Clouds, rain, wind or fog could impede their work, and the chances of having good weather in all eight locations simultaneously was low. There was another factor that could impact the observations: the human one. The telescopes were all at high altitude, and the low oxygen levels were punishing, taking their toll on the scientists' bodies, short-term memory and concentration.

While Sara Issaoun was installing herself at the top of Mount Graham, Heino Falcke was preparing for observations at the IRAM 30-meter telescope in the Sierra Nevada mountain range in southern Spain.[15] The observations were being coordinated globally by Shep Doeleman and his colleagues from a campaign centre at the offices of the Black Hole Initiative in Cambridge,

THE SHADOW HUNTERS

Massachusetts. The Black Hole Initiative is a research centre at Harvard University that takes an interdisciplinary approach to the study of black holes. On a whiteboard in one of the organization's offices, Doeleman could make notes on the status of the telescopes. Their success depended primarily on one factor: the weather, since there were only a handful of days each year on which the conditions would be good enough to see Sagittarius A* simultaneously with all the telescopes. Time was tight.

Observations began on 5 April 2017. As well as Sagittarius A* and M87*, the EHT was observing a number of other black holes. Even if it wasn't possible to see their shadows, the team could gain important information from them.

'I kept thinking about how the telescope was calibrated and what the weather was like,' Issaoun says of her first shift. After working for thirty hours straight she tried to get a little sleep. She crawled into a bed in the lower part of the facility, but was disturbed by the constant noise of the telescope rotating above her bedroom. After a few hours of interrupted sleep it was time to start her next fourteen-hour shift.[16] The data collection continued for ten long days.

NEW IMAGE ALGORITHMS

'We're done!' exclaimed Shep Doeleman on the tenth day. 'It's a wrap.'[17]

He put on the song 'Somewhere Over the Rainbow', singing along to the line 'Dreams really do come true'. On the EHT group chat he wrote that the drinks were on him that evening. The astronomers at the telescopes could finally let out the breath they'd been holding. They'd pushed themselves to the limit. 'I felt like I would never want to do it again,' Issaoun recalls. 'I was just happy to go home.'

Now it was time for two supercomputers in the US and Germany to analyse the material to see if it was possible to join up the telescopes' data into a combined signal. In total they had gathered 4,000 terabytes of data, stored on more than half a tonne of hard drives that were spread out across the different sites. Sending such huge volumes of data over the internet would take too long, so the disks had to be transported over land, air and sea. Since half a tonne of hard drives is very expensive, the astronomers couldn't afford to back up all the data. Little wonder, then, that there was a sign in one of the data centres that said: 'ATTENTION!!! Hard drives are more sensitive than eggs.'

For several months the two supercomputers chewed over the contents of the disks. They were looking for a signal that would allow them to link the data from all the telescopes into one gigantic virtual telescope. In order to extract the signal they needed the telescopes' observations to be coordinated in space and time. Advanced atomic clocks in the telescopes recorded the exact moment at which each radio wave reached them. The astronomers also needed to keep track of how the movements of the tectonic plates, the Earth's rotation and other subterranean movements affected the telescopes' positions on the Earth's surface. It was a challenging measurement, and the astronomers would only be able to find out whether it was possible to reconstruct signals from one of the universe's most extreme places once all the data had been joined up in the supercomputers.

The data analysis worked. They were able to correlate all the data from the telescopes. In October 2017 several of the members of the EHT met at the offices of the Black Hole Initiative to take the analysis to the next stage. They needed to develop new image algorithms so that they could turn the signals the

THE SHADOW HUNTERS

supercomputers had extracted from the huge mass of data into a 64×64 pixel image of what happens around a black hole. Shep Doeleman welcomed the participants and reminded them that, just like Jean-Pierre Luminet, they were pioneers in the exploration of black holes.

The biggest problem the image analysis group faced was that, because they had used eight networked telescopes instead of one big one, it wasn't obvious how they should translate all the data into a single image. 'There were an infinite number of images that matched the data,' Issaoun explains.

Imagine, for example, that you are going to take a photo of someone's face. Instead of one camera that can photograph the whole face, you only have access to several small cameras that can take pictures of different parts of the face. With these images, which might depict part of one ear, an eye, a little of the hair and so on, you have to reconstruct the whole face. There might be several faces that would fit with the details you've photographed.

Similarly, the EHT's members had to reconstruct an image from a limited amount of information. But there were two key differences to taking a picture of a face. First, the radio telescopes did not collect data in the form of pixels in an image, but in the form of abstract visual components that could be translated into images in a number of ways. The technical term for this kind of translation is Fourier analysis, and it is fundamental to all kinds of image and sound processing. Secondly, they had no precedent with which to compare the images they extracted. When you reconstruct an image of a face you can compare the result not only with the face you have photographed, but with all the faces you've ever seen. But no one had ever seen or depicted a black hole before, so the only comparisons they could make were with data simulations of how it should look. Therein lay the greatest danger: 'As humans, we sometimes see what we want to see,' says

Issaoun. 'That's why we as scientists need to explore all options before we can draw a solid conclusion.'

A few years previously, a group of astronomers from the US had made a scientific blunder. They had told journalists at a press conference that they had detected signals from the initial moments of the universe itself, with the help of the telescope BICEP2 at the South Pole.[18] There was a flurry of speculation about a Nobel Prize, but more detailed analysis showed that the signals came not from the beginning of the universe, but from dust in our own galaxy.[19] The research group withdrew their conclusions, and the incident became a cautionary tale about the dangers of presenting a result to the media before a thorough scientific analysis has been conducted.

Heeding this lesson, Sara Issaoun and her colleagues divided themselves into four separate working groups. Only if all the groups produced the same results would they be able to trust their image algorithms.[20] With the aid of data simulations, they produced an archive of all the possible ways a black hole could look. The simulations took into account how light travels through the complex interplay of magnetic fields, swirling plasma and spacetime curvature that surround a black hole. Then they simulated how the telescopes would have observed this light and whether their algorithms were able to accurately reconstruct the black hole image from the sparse data provided by the telescopes.

They decided to attempt an image of M87* first. There was a practical reason for this: as the Earth rotated, the EHT telescopes had gathered different kinds of information from the black hole they were observing at that time. The observations for a whole night could then be assembled. For M87* this posed no problems, as the matter moving around it takes up to a week to complete one orbit, but with Sagittarius A*, which is much smaller than M87*, matter completes an orbit in anywhere from

THE SHADOW HUNTERS

a few minutes to an hour. This meant that when the night's observations were overlaid, it was like using a long exposure time to photograph someone running – the image would be completely blurred. By starting with M87*, the astronomers were stripping out this layer of complexity.

After developing and testing their algorithms, the members of Issaoun's group were ready to make their first attempt at composing an image from the telescope's data. Everybody in her group would undertake this process independently. 'We wanted everyone to have their own personal experience of creating the first image of a black hole,' she says. 'It was a once-in-a-lifetime opportunity.' Alone in her office, Issaoun wrote a program to load the data, run the group's algorithm and then display the image. After a few minutes a black hole started to take shape on the screen. The image was full of noise but Issaoun could clearly see that it contained a ring of light around a dark central area. 'I was so excited when I saw it,' she recalls.

At the end of July 2018 the four groups met in the Black Hole Initiative's conference room to see if they'd produced the same result. They transferred their images of M87* from password-protected folders so they could look at them together. When they saw all the images they were captivated. Every single one showed a dark area surrounded by a bright ring that was lighter at the bottom than the top. 'It was really exciting to finally see the shadow staring back at us,' says Issaoun. 'The fact that we actually see the shadow gives us pretty definite evidence that this object is a black hole.'

Around the black hole there are gas particles whirling at enormous speeds, crashing into each other in what Shep Doeleman calls 'a cosmic traffic jam'. The fact that part of the gas cloud looks brighter on one side is due to the Doppler effect that Luminet had noticed in his 1979 drawing.

156 FACING INFINITY

Now that the image had been created, it was possible to calculate the size of M87*. Astronomers use degrees to measure how large objects appear in the sky, with the distance across the dome of the sky from horizon to horizon representing 180 degrees. One degree can be divided into sixty parts, called arcminutes, and an arcminute can, in turn, be divided into sixty arcseconds. The Moon and the Sun are each around thirty-one arcminutes across, while the star Betelgeuse is only around 0.05 arcseconds. The shadow of M87* is much smaller than that: forty-two thousandths of a thousandth of an arcsecond. This measurement, in combination with the known distance to M87*, makes it possible to calculate the actual size of M87*. Its diameter was found to be thirty-eight billion kilometres – larger than Pluto's orbit around the Sun – and scientists were then able to use Schwarzschild's formula to determine the mass of the black hole at 6.5 billion solar masses.[21] All this was a brilliant reminder of how far-reaching humanity's mathematical abilities are: even though the black hole is fifty-five million light years away, the EHT scientists were able to study it in depth.

That evening, Issaoun and her colleagues went to a bar in Boston to celebrate their work on M87*. They moved the tables and sat in such a way that the table and the scientists around it looked like the image they had just created. They did karaoke, with someone choosing the song 'Black Hole Sun' by Soundgarden. As they sang along they changed the 'sun' to 'shadow', and so the other people in the bar were treated to a group of astronomers bawling out, 'Black hole shadow, won't you come'. The members of the four teams knew they had witnessed something that no one else had ever seen. 'It was the best day of this project,' Issaoun says.

But the next day they were back at work, doing more analysis. They needed to be completely sure they weren't fooling

THE SHADOW HUNTERS

themselves, seeing what they wanted to see. 'We were excited about the image but we were showing a poker face on the outside,' Heino Falcke tells me. 'That's the emotional state of scientists who are on the verge of making a discovery.'

As well as conducting more tests, the members of the EHT had to quickly turn out six academic papers and prepare a global press conference. The workload was enormous. In the weeks before the press conference Issaoun and her colleagues were working sixteen-hour days, every day of the week. As the work piled on, friction within the team grew. 'We were like a steam pot just waiting to explode,' says Issaoun. 'If there are conflicts and you have time to resolve them, then it's fine. But if you don't, the tension builds up. People had completely random outbursts. Some would get angry for no apparent reason, it just happened to be something at the time that pushed them over the edge. There was a lot of stress.'

The stress was intensified by the secrecy. They had produced several images of M87* from across the observation days, and they had to make sure none of them got leaked to the press. Even though only EHT members and a handful of communications teams of various institutes were allowed to see them, there were concerns that they would get out.

To an outsider, the Event Horizon Telescope might seem like a highly ambitious and impressive scientific collaboration. But when I speak to the organization's members I can tell that many of them have been affected by the internal conflicts. They are reluctant to go into detail or name specific individuals, and seem keen to put the conflicts behind them, but it's clear that a research project across multiple continents, in which huge amounts of prestige – and money – are at stake, takes a mental toll on the scientists involved. They were pushing both the technological resources and themselves to the limit, in order to

158 FACING INFINITY

depict something that's almost impossible to see. Their scientific ambitions came at a human price.

THE FIRST IMAGE IS RELEASED

On 10 April 2019 the Event Horizon Telescope team were ready to present the first image of the black hole M87* to the world.[22] The image was shared simultaneously at press conferences in Washington DC, Brussels, Santiago, Shanghai, Taipei and Tokyo. 'We have seen what we thought was unseeable,' said Shep Doeleman at the press conference in Washington DC. 'We have seen and taken a picture of a black hole.'

'At the huge distance of M87* it actually appears as a mustard seed in Washington DC as seen from here,' Heino Falcke told a press conference in Brussels, concluding with a line that felt like something out of his former life as a lay preacher: 'It feels like we are looking at the gates of hell at the end of space and time.'

After the image of M87* (see page 162) had been presented, it spread rapidly.[23] It may have taken light leaving the vicinity of the black hole fifty-five million years to reach Earth, and two years of analysis to produce an image, but after the press conference it was a matter of minutes before the image had been seen by many people around the globe. M87* was splashed across the front pages of most major newspapers. A media analysis estimated that several billion people had seen the image within the space of a few days. 'It's rare to have science news on front pages around the world,' says Issaoun. 'Through our work, people could feel a sense of unity.'

The fact that the image got so much coverage is due partly to the associations black holes spark in the imagination. The phrase itself – 'the shadow of a black hole' – gave the image a poetic feel. 'I think it's the right word because it also has a mythological meaning,' says Falcke when I ask him about the choice of words. 'It's not just a scientific term, just as black holes or the Big Bang are not. The term has multiple layers which are not just scientific, but which appeal to the public.'

But someone floating close to M87* wouldn't see the orange ring we see on the EHT image. The telescope observes radio waves with a wavelength of 1.3 millimetres, more than two thousand times longer than what our eyes can detect. The image Issaoun and her colleagues created shows how the intensity of the radio waves of this precise wavelength varies around the black hole. The image could have been greyscale, with black representing the weakest light intensity and white the strongest, but when Heino Falcke published simulated images in 2000 that predicted what the black hole shadow would look like, he had chosen the red–orange colour range. 'I tried all the colour maps,' he says, 'and there was one called *heat*. I thought the black hole is a hot and hellish place. So I used those colours.' There is a problem with associating red and orange with heat, however. In a flame, the hotter interior of the flame is blue in colour, while the outside is redder. 'We discussed the choice of colour,' says Falcke, 'and although blue corresponds to the hottest part of a flame, we felt that orange conveyed heat more effectively and immediately.'

After the image of M87* was released, the Museum of Modern Art in New York and the Rijksmuseum in Amsterdam both decided to include it in their photographic collections. The status of the image went beyond that of a scientific artefact and raised questions about what we even think of as a photograph. Black holes are the most extreme objects in the universe, where matter has been compressed to the point of unrecognizability, their edges a place where space and time themselves form a limit to our knowledge. In order to depict what happens at that limit, scientists had to push their methods of observation to the absolute extreme. They had to link up telescopes around the world, put supercomputers to work handling all the data, and develop new image algorithms to make the invisible visible. But the last step in this process is not about the technology but about us. At the boundary of knowledge we must also challenge our visual capabilities. The images of black holes force us to think critically about how we make the world accessible to our senses when we try to see what is happening at the limits of the visible.

The members of the EHT were therefore disappointed when some people called their image 'blurry'. 'When people say that, I am surprised,' Sara Issaoun says. 'The image of the shadow is the highest-resolution astronomical image ever made. The Event Horizon Telescope can see three million times better than our eyes. The image is blurry because the black holes are very far away.' Films like *Interstellar* and the dazzling animations in TV documentaries have led us to expect sharp, detailed images of objects in space. But for Sara Issaoun, the picture of M87* is just the first step. 'If you think about other first images in the history of humanity, they have always been a bit blurry.'

When I talk to Shep Doeleman, who worked for over twenty years to realize his vision of depicting black holes, he takes a historic perspective on the image. 'When people made maps of the globe in the past, it would show dragons,' he says. 'They were a warning saying, "Don't go there, we don't know what is happening there." Thanks to the image of M87*, we know that there are parts of the universe that are beyond our reach. That is scary. Western philosophy is predicated on the idea that if you know everything, now you can predict everything that is going to happen in the future. But if you fall into a black hole, there's no way of predicting what's going to happen.'

THE SHADOW HUNTERS
161

Heino Falcke also mentions the fact that their image of the black hole represents a limit to our knowledge. 'In a way, the image is a huge success,' he says, 'but you can also consider it a huge disappointment. Now we see the darkness of the event horizon. That's why I sometimes call it our final battle.' There is another sting to the picture: it reminds Falcke of the difficult conflicts that accompanied its creation. 'I also see the pain behind the picture,' he says. 'The human battle was bigger than the scientific battle.'

In one of their papers, the EHT scientists wrote that the observed image was consistent with the predictions Roy Kerr's mathematical formula made about the shape of the shadow (or, as Falcke put it: 'It walks like a black hole, it quacks like a black hole, so it should be a black hole.')[24] When I interviewed Kerr I asked him what he thought of the image. 'It's amazing,' he replied. 'If I was dishing out Nobel prizes, I'd give one to that group.'

Jean-Pierre Luminet, who had published his illustration of a black hole four decades earlier, was travelling in Uzbekistan when the image was presented. At the same time the EHT was holding its press conference, he was giving a presentation to the Uzbek Academy of Sciences in the capital Tashkent. While the academy's members went out for lunch, Luminet ran up to his hotel room and watched over a shaky internet connection as Shep Doeleman showed the image of M87*. 'I never thought it would be possible to see it in my lifetime,' reflects Luminet. 'The image plays a fundamental role in our understanding of the real structure of space and time.'

The first time I saw the picture it sent a shiver down my spine. It was proof that up there among the stars, planets, moons and galaxies we love to gaze upon, there is also a little area of the sky that is completely dark. No light can escape it, and so it constitutes an absolute limit to our knowledge of the universe. I was also deeply impressed by the fact that these scientists had managed to study the darkness of a black hole that is fifty-five million light years away in another galaxy. I keep trying to come up with a way to formulate what an incredible achievement that is, but because no one has ever done anything like it, it's hard to find an appropriate comparison. I decide it's enough to know that the image changed my life, because it made me start writing this book.

A NEW BLACK HOLE IMAGE

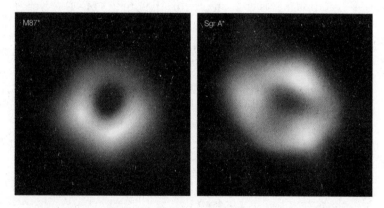

The two black holes imaged by the Event Horizon Telescope Collaboration. M87* lies at the centre of the galaxy M87, around fifty-five million light years from Earth. Sagittarius A* lies at the centre of the Milky Way.

'This is the first image of the supermassive black hole at the heart of our Milky Way galaxy, Sagittarius A*,' said Sara Issaoun proudly at a 2022 press conference at the European Southern Observatory's headquarters near Munich. After several years of complex analysis, she and her colleagues had finally managed to produce an image of the black hole at the centre of the Milky Way.

The EHT's members used the hashtag #OurBlackHole when talking about the new image on social media. They wanted to emphasize the fact that this was the black hole we are travelling around on our 230-million-year journey around the Milky Way. 'For decades,' Issaoun continued at the press conference, 'we have known about a compact object that is at the heart of our galaxy that is four million times more massive than our Sun. Today, right at this moment, we have direct evidence that this object is a black hole.' When I meet Issaoun, she tells me how nervous she was during that press conference. As with M87*, the picture of Sagittarius A* would be broadcast simultaneously at multiple coordinated press conferences. Once again, there was great secrecy.

The pictures of Sagittarius A* and M87* both showed a pale ring surrounding an area of darkness. The size of the two shadows was almost the same. 'When you see the images of Sagittarius A* and M87*, they look quite similar,' Sara says. 'I love that, because black holes are such beautifully simple objects. They were predicted mathematically by Karl Schwarzschild. Einstein himself thought that it's a beautiful mathematical thing but that such a thing could not possibly exist. How could something so simple and yet so abstract be out there in the universe? But we see the shadow and the ring, which are a pure signature of strong gravity, of spacetime getting bent. In the centre of the image you can see the end of space and time. That's crazy to me.'

Both Andrea Ghez and Reinhard Genzel had studied the stars orbiting Sagittarius A* to confirm that its mass was equivalent to just over four million solar masses. Early measurements of the motion of dust clouds at the centre of the Milky Way had indicated a similar mass. The EHT image had now provided another way of calculating the object's mass: measuring the size of its shadow. The fact that each of these methods produced the same result was important for confirming that Sagittarius A* really was a black hole.

Sara Issaoun's astronomy journey started when she heard Andrea Ghez give a lecture on Sagittarius A* as a student. Several years later, she herself was presenting the first image of it. Her journey to that point started with an email she sent to Heino Falcke. 'It was just chance that got me into this project,' she reflects. 'Often when I give public talks, young people who want to become astronomers ask me how I managed to become part of the Event Horizon Telescope. The only advice I can give is to just ask. You have to grab the opportunities as they come.'

JETS AND MAGNETIC FIELDS

The images of M87* and Sagittarius A* are useful for more than just finding out the size and mass of the black holes. One of the most important questions the EHT researchers wanted to answer with these images was how black holes can give rise to one of the brightest lights in the universe. It sounds like a contradiction: black holes are meant to create darkness, not light. But in 1918 the US astronomer Herbert Curtis discovered a 'curious straight ray' shooting out of the galaxy M87. It was 5,000 light years in length, and when astronomers studied it with radio telescopes they could see that it extended even further, up to 100,000 light years out of the galaxy. That makes the jet, as these extended objects came to be called, almost as long as the Milky Way is wide.[25]

A jet consists of charged particles travelling out of the centre of the galaxy at almost the speed of light. Just as a jet of water becomes more focused the narrower the mouthpiece of the hose, the structure of this galactic jet suggested it must have been produced from a tiny volume. It was like a gigantic finger pointing back at the centre of the galaxy and saying, 'Look here, this must be an area where extremely strong forces operate.'

The galaxy Hercules A can be seen as a fuzzy blob at the centre of this image. Two jets are shooting out of the galaxy, and turn into extended lobes further out. The jets are created by a supermassive black hole at the centre of Hercules A, and span more than a million light years. Similar jets have been observed in many other galaxies, including M87. The image has been created from observations by the Hubble Space Telescope and the Karl G. Jansky Very Large Array radio telescope in New Mexico.

Similar jets have been observed in other galaxies. Their vast energy is generated by the powerful gravity around supermassive black holes. Despite the fact that black holes draw in matter that is close to them, surprisingly they can also do the opposite. A complex interaction between powerful magnetic fields, the rotation of the black hole, and the matter that orbits it means that not all matter travels *into* it. Similar to the way a baby eats and splatters food in all directions, matter can travel away from the black hole, and some of this is channelled into these enormous jets.

It is still unclear exactly how these jets are formed, and in order to gain more clarity, astronomers will need to study what is happening up close to a black hole. With the aid of an updated version of the image of M87*, the EHT researchers were able to show that the magnetic fields close to the event horizon were playing an important role in the creation of the jet.

In 2023 the network Global Millimeter Very Long Baseline Array presented an image of M87* in which both the shadow and the base of the jet could be seen simultaneously.[26] In 2024 the EHT researchers were able to show how M87* had changed in the year following their initial observations.[27] The ring had rotated by thirty degrees, probably because of the turbulent movement of the gas swirling around the black hole.[28]

Astronomers have as yet been unable to observe a jet coming out of Sagittarius A*. It's possible that it does exist, but if so, it must be very weak. There's another peculiar facet to this black hole: it appears to be on a diet. The EHT scientists have compared the black hole's 'eating habits' to a person who consumes only half a mouthful of rice every millionth year. Sagittarius A* may have once been an active black hole that swallowed up huge quantities of matter and expelled a powerful jet, but today it comes across as a rather unassuming old cosmic giant.

Like M87*, Sagittarius A* is surrounded by a powerful magnetic web. 'We have learnt that strong and structured magnetic fields are crucial for how black holes interact with the gas and matter around them,' said Sara Issaoun when a new image of Sagittarius A* was released in 2024.[29]

Over three hundred scientists from Europe, Asia, Africa and North and South America have worked on producing and updating the images of these two black holes' shadows. The scientific value of the images may be great, but the significance of their mere existence may be even greater. They have enabled us to

THE SHADOW HUNTERS

form a relationship with these two supermassive black holes, to gaze into their darkness and follow what happens to them over the years to come. 'We feel a sense of ownership because it is the black hole at the centre of our own galaxy,' says Issaoun of the image of Sagittarius A*. 'But the image isn't the end for us, it's more of a beginning.'

An updated version of the image of M87* was released in 2023.[30] It showed a much sharper ring, which was made possible by machine learning algorithms the researchers used to analyse the telescope data. But the next big step is not about sharper images, but actual movies of the black holes. These would show how the matter around them changes over time. One of the key aims of the new observations is to investigate whether Einstein's general theory of relativity correctly describes how black holes work, or if there are new physical phenomena that would be better described by a different theory.

How the observations might be improved is one of the points of discussion at the conference in Granada where I meet Sara Issaoun. One way is to use more telescopes on Earth. Heino Falcke and other scientists want to build a telescope in Namibia called the Africa Millimetre Telescope. Shep Doeleman wants to build several new telescopes around the globe as part of the project Next Generation Event Horizon Telescope. One planned site lies on Tenerife in the Canary Islands, where the Tenerife Event-horizon Antenna will be situated. Another possibility is upgrading the radio telescopes already in the network, so that shorter wavelengths could be observed, resulting in both sharper images and the ability to see light emissions at different wavelengths. A first step in this direction was reported in August 2024.[31]

Looking ahead, scientists are also contemplating the exciting prospect of putting telescopes into space. 'There are two ways to get sharper images,' Issaoun says, 'make the telescope bigger or go to shorter wavelengths. If you have space telescopes, you can do both.' This is a major challenge, however. 'We do not yet have the technology required to place radio dishes in space that can observe at such short wavelengths,' she continues. 'We collect petabytes of data, and sending that back down to Earth with very high data rates is our biggest hurdle.' In spite of the barriers, she is optimistic. New technology that can send data from space via laser beams could make it possible to reach the high bandwidth the EHT telescopes require, even if the idea of signals from black holes being sent down to Earth with laser beams sounds like science fiction. 'The Event Horizon Telescope has been on the edge of technological developments during its entire history,' concludes Issaoun. 'We will continue with that until we finally achieve the breakthrough that puts the telescopes in space which can give us amazing sharp images and movies.'

The story of the images of M87* and Sagittarius A* is thus a story of the very human determination to overcome our limitations. Creating images of something regarded as unseeable requires new technology and collaboration across multiple continents. Ultimately, the scientists of the Event Horizon Telescope were able to create a virtual telescope spanning the whole world, in which the Earth's rotation augmented the technology that made it possible to depict supermassive black holes.

But the images of M87* and Sagittarius A* also raise a question: where do these supermassive black holes actually come from?

Chapter 9

· · · · · · · · · · · · · · · ·

THE ORIGIN
OF THE GIANTS

It's September 2021, and I emerge from the Denfert-Rochereau metro station into the warm, late-morning Parisian sunshine. The hectic traffic rushes past me, and I look beyond it to the entry to the catacombs on the other side of the avenue. They were built in the eighteenth century, when the mayor of Paris decided to relocate the contents of the local cemeteries to a series of disused quarries and mines. The bones of centuries of dead Parisians were piled up in the underground passages, and eventually the millions of skeletons became a tourist attraction.

I walk past the entrance to the catacombs and continue along Boulevard Arago. I'm on my way to a meeting where I'm hoping to learn more about how supermassive black holes come into existence. Sagittarius A* and M87* are the most famous supermassive black holes, but astronomers have identified several million others. They occupy the centres of most galaxies and the mass of some is equal to that of many billion Suns.

Supermassive black holes seem to have a close relationship to the galaxies they are found in. The galaxies offer a good stock of gas clouds to make them grow, and in turn the black hole jets can influence the evolution of the galaxies. Everything is connected: the mass of the supermassive black holes and the galaxies, the number of stars formed and how fast they move. This has created something of a chicken-and-egg situation: what came first, the galaxy or the supermassive black hole? If astronomers can answer this question, they will gain a better understanding of how different structures in our universe form and develop, and, by extension, how our own galactic home, the Milky Way, came into being.

But while we can study the past of Paris through its street names, museums, wall plaques and even the catacombs, it's not possible to tell where black holes come from just by looking at them. Since no information can escape from them, it is impossible to know how they were created and what has since fallen into them; black holes erase the memory of their own origin. Astronomers must therefore undertake cosmic detective work to solve the mystery of their creation.

I turn off Boulevard Arago, pass through a tall iron gate and enter a garden. Through some trees I can see the shining white domes of the Paris Observatory outlined against the blue sky. I find Marta Volonteri standing by a park bench. She is one of the world's foremost experts in the formation and development of supermassive black holes. Her workplace, the Institut d'Astrophysique de Paris, is a stone's throw from the gardens of the observatory.

'As you can probably tell from my accent, I'm from Italy,' Volonteri says with a laugh when I ask about her own origins. She tells me she has done research on black holes at institutes in various countries: twice in the US, then in the UK and now in France. 'I've forgotten what it's like to live in my home country,' she says, a note of resignation in her voice.

'Are you a nomad?' I ask.

Her answer hints at the vagaries of the academic career path: 'No, I'm an astrophysicist.'

THE ORIGIN OF THE GIANTS

My mind returns to something Volonteri once wrote in the afterword to an academic paper. She said that she was 'grateful to all my more recent collaborators for putting up with my sometimes insane work rhythms. I promise I am a much more relaxed and pleasant person when I sit on a sandy beach with a book in my hands, the Sun shining, and the ocean glistening.'[1]

To me, Marta Volonteri comes across as very relaxed. She smiles a lot as we talk, but behind the smile lurks a very serious intention: to understand the origins of the biggest and darkest objects in the universe. For more than twenty years she has been wrestling with the question of how the universe came to be riddled with supermassive black holes.

'My research has gone from being science fiction to becoming science,' Volonteri says. 'I had to fight for my research for many years, because people wouldn't believe in it. When I started, we hadn't even imaged black holes or seen gravitational waves from them, so people asked me why I was wasting my time. But now it's all reality.'

Volonteri tells me that, according to the basic model of the genesis of supermassive black holes, a star or gas cloud collapsed into a black hole at some point in the universe's infancy. The black hole grew larger and larger as it sucked in matter from its surroundings, ultimately becoming one of the giants we see today. 'But the more physics we add to this model,' she explains, 'the more difficult it becomes to understand.'

Scientists' uncertainty about the origins of black holes is due mainly to the fact that they turn up so early in the history of the universe. The universe is approximately 13.8 billion years old. Astronomers have observed supermassive black holes that appear to have existed less than a billion years after the Big Bang (a term for the extremely dense and hot initial moment of the universe), but have a weight of more than a billion solar masses.

172   FACING INFINITY

That is a lot of mass to accrue in such a relatively short time. It's like looking at a photo of someone from their time in nursery school, but instead of seeing a young child's body there's a fully grown person staring back at you.

Volonteri says that in order to explain this rapid growth she and her colleagues need to know what 'seed' the supermassive black hole grew from. Astronomers have long known of one such potential seed: stars.[2] The first stars in the universe were formed at some point in the first hundred million years after the Big Bang. They lit up the darkness of the universe and came together in the gigantic stellar cities we call galaxies. The stars consisted of nothing but the three elements that were available in the aftermath of the Big Bang: hydrogen, helium and a tiny quantity of lithium. Because of the interaction between the light the stars generated and the elements they consisted of, these primordial stars were probably very large, short-lived but burning with an intense brightness, like the mayflies of space.

Some of the first stars in the universe may have imploded into black holes. These black holes were themselves a kind of seed, with a mass equivalent to several hundred Suns.[3] The more they ate of the cosmic buffet around them, which consisted mainly of clouds of gas, the more they grew. 'But it seems increasingly unlikely,' Volonteri tells me, 'that black holes of billions of solar masses can be created in the first billion years of the universe starting from a seed created by a normal star.' The reason is that a black hole regulates its own consumption. Gas that falls into it starts to glow because it crashes into and whirls around other gas particles. This radiation creates a pressure that stops more gas reaching the black hole.[4] The more the black hole consumes, the harder it gets for new material to fall into it, setting a limit to the speed at which black holes can grow. Another issue is that black holes that formed from collapsing stars are relatively small,

THE ORIGIN OF THE GIANTS

notably smaller than the stars that made them. 'The gas really has to hit the black hole with high accuracy to be swallowed by it,' says Volonteri, 'and it has to happen continuously. Although it's not impossible, it's very unlikely.'

A reasonable assumption is therefore that supermassive black holes started out with greater mass than is possible from the collapse of an ordinary star. One possibility is that black holes were formed from the collapse of gigantic gas clouds. Seeds of this kind are known as direct collapse black holes, and from the start they can have a mass of several hundred thousand Suns. One version of this formation theory states that before the gas clouds became black holes, they underwent a short phase as supermassive stars. These are a kind of hypothetical giant star that might have been several hundred thousand times the size of our Sun. They could only have existed under the unique conditions that prevailed at the dawn of the universe, so we do not see them in today's cosmic environment. Black holes that were formed from collapsing gas clouds, and, possibly, supermassive stars, can have a huge head-start on their counterparts born from ordinary stars. However, the conditions have to be just right for direct collapse black holes to occur. 'It can happen,' Volonteri says, 'but nobody thinks it's common.' To figure out what the seeds of supermassive black holes are, it is therefore crucial to understand how common it is for gas clouds to collapse directly into black holes, as well as to know whether the extreme supermassive stars could have been present in the early universe.

THE BRIGHTEST OBJECTS IN THE UNIVERSE

The traffic noise disturbs our conversation, so Volonteri and I get up and walk further into the observatory gardens. When I see the domes I think of Pierre-Simon Laplace, who in the late eighteenth century advanced the theory that the majority of objects in the universe might be dark and invisible. Laplace was a member of the Bureau des Longitudes, which ran the Paris Observatory.[5] Perhaps he looked out of its enormous windows at some point and reflected on the possibility that the largest objects in the universe do not emit light. Both Laplace and Britain's John Michell had suggested that these dark stars might have a mass equivalent to more than a hundred million Suns. As previously mentioned, the idea of these objects was rooted in Newton's theory of gravity, and as such was very different from the modern view of black holes as created from spacetime curvature. But it is fascinating that Laplace and Michell were considering the possibility of such gigantic dark celestial bodies so long ago.

The first empirical sign of the existence of black holes was found in the early sixties. At the same Texas conference where Roy Kerr presented his formula for rotating black holes, delegates were discussing a surprising astronomical discovery. The Dutch astronomer Maarten Schmidt had demonstrated some puzzling characteristics in an object named 3C 273.[6] The object looked like a star – a shining point in the sky that appeared to be located in the Milky Way – but when Schmidt studied the light in detail he confirmed that it wasn't a star in our own galaxy, but an object located more than two billion light years away. 3C 273 emitted as much light as several trillion Suns, much stronger than the brightest galaxies then known. At the same time, small variations in the radiation from 3C 273 indicated that whatever it was that was releasing this vast quantity of energy, it was doing so from a very small volume, less than a thousandth of the size of our galaxy.

3C 273 and other similar objects came to be known as *quasars* (an abbreviation of *quasi-stellar*). As the name suggests, quasars

THE ORIGIN OF THE GIANTS

looked like stars in the sky, but they were actually something else. 'No one knew what a quasar was,' Kerr told me when I interviewed him, 'except that you shouldn't be near one.'

Scientists speculated that the quasar radiation was generated by matter falling into and spinning around gigantic black holes.[7] Just as a spaceship starts to glow as it enters the Earth's atmosphere and collides with the particles in it, matter orbiting the black hole would glow as particles crashed into one another at enormous velocities. The gravity around a black hole is so strong that they are believed to be the universe's most efficient converter of matter into energy and radiation, significantly more efficient than the fusion in stars or the fission inside a nuclear reactor. A black hole's rotation can also contribute to the generation of energy. Through this interplay between gravity and matter, quasars are supermassive black holes that release copious amounts of energy into the universe. In spite of their darkness, black holes can therefore create huge quantities of light, and many quasars also emit the type of jets we encountered in the previous chapter. It is undeniably a cosmic irony that supermassive black holes give rise to the brightest light in the whole universe.

The energy they emit affects how many stars are created in the galaxies in which they are found (which we will discuss further in Chapter 12). Black holes can therefore be both destructive, in that they swallow matter, and creative in that they help shape the universe's cosmic structures. 'I think of some gods in ancient mythologies who are both creators and bringers of light and joy, but can also bring destruction. Black holes have some of those aspects,' Volonteri reflects.

Today astronomers have observed more than a million quasars. In spring 2024 the German astronomer Christian Wolf and his colleagues reported that they had discovered the brightest quasar in the universe, which they named J0529-4351.[8] This majestic quasar has a mass equivalent to seventeen billion Suns, gobbles a Sun's worth of matter every day and is 500 trillion times more luminous than the Sun. Another quasar does not shine as brightly but nevertheless has a larger mass. The hulking Tonantzintla 618 is ten billion light years away and weighs forty billion solar masses (as you can see, 'billion' is a word frequently associated with supermassive black holes!). The quasar was named after the city of Tonantzintla in Mexico, the site of the telescope with which it was discovered. Top candidate for the title of largest black hole ever observed is called Phoenix A*, weighing in at an estimated one hundred billion solar masses.

Quasars have not been equally active throughout the universe's existence. According to astronomers, it seems they were most common around 2.5 billion years after the Big Bang. When ready access to matter dried up, the party was over and gigantic black holes entered a calmer phase. That's why we no longer see any active quasars in the present-day universe. Not all black holes were once a shining titan of this kind either. Perhaps some spent their days quietly in the centre of a galaxy, without access to the gas and matter that would make them grow to vast sizes and shine brightly while launching jets into the universe.

But the most significant discovery to emerge from the observations of the quasars was that supermassive black holes existed so early in the history of the universe.[9] The question remains: where do supermassive black holes come from?

DARK MATTER

In the shade of some trees Volonteri and I find a bench where we can continue our conversation. She tells me that there is another possible seed. In the very earliest moments of the universe neither

THE ORIGIN OF THE GIANTS

stars nor atoms had formed. Instead, space was filled with subatomic particles, light and, possibly, remnants of other energy forms whose densities could have fluctuated strongly across different regions. 'Then there exist little pockets of the universe that are so dense they could collapse into black holes,' she says.

These hypothetical black holes are known as primordial black holes. The possibility of their existence was suggested by a group of Soviet physicists in 1967.[10] Since then they have been studied both theoretically and searched for observationally by hundreds of scientists. If they exist, they could provide an answer to the galactic chicken-and-egg problem: first came the primordial black holes, they grew larger and larger and then galaxies formed around them.

One appealing idea is that the primordial black holes might explain the cosmic phenomenon of *dark matter*. Astronomers have found signs of this strange phenomenon in a range of observations: from the cosmic background radiation to the movement of the stars in galaxies, and from the clumping of galaxies into large-scale structures to the way light travels through the universe. It seems that something is generating an even stronger gravitational force than should be possible from the ordinary matter that we can see with the help of telescopes. This dark matter does not emit light and can therefore not be observed directly, but it does seem to gather into lumps and, through its gravity, it affects how structures in the universe form and evolve.

While astronomers have been searching for other kinds of signals from this matter in space, physicists have been searching for it in underground detectors. Although the hunt for dark matter has been going on for many years, a clear signal has yet to be observed.

But there is no lack of ideas about what dark matter might be. A common hypothesis is that dark matter consists of one or more kinds of particles that do not interact with light. But the complete invisibility of dark matter has led some, quite justifiably, to ask: is it not possible for black holes to be part of this dark matter? It certainly feels like a reasonable suggestion. After all, black holes are dark and they do not emit any light. However, when astronomers study the cosmic background radiation, which was created when the universe was a mere 380,000 years old, they find that dark matter appears to have existed in the universe from very early on, since long before the first stars and galaxies existed. Black holes that came into being later – through the collapse of stars and clouds of dust – cannot therefore be the main contributor to dark matter. What's more, the hundred million or so black holes that have been formed from the implosion of stars in our galaxy do not jointly have enough mass to constitute all the dark matter that is spread in a halo around the Milky Way.

The possibility remains, though, that primordial black holes represent part of, or all, the dark matter in the universe. Stephen Hawking suggested back in 1971 that primordial black holes might constitute a large part of the mass of the universe as a whole.[11] If they *were* created, they still exist today (as long as they haven't disappeared in the way Hawking predicted they might, something we'll come to in Chapter 13). Astronomers have long searched for signs that such black holes exist, but thus far, no decisive signal has been observed.

So we're still lacking conclusive proof that black holes are part of this dark matter, but it hasn't been completely ruled out. If future observations were to show that primordial black holes make up either part of or all the dark matter in the universe, or that they are the seeds of supermassive black holes, it would undeniably bring a new dimension to their importance in the evolution of the universe. These primordial black holes from

THE ORIGIN OF THE GIANTS

the universe's earliest days could have influenced how some of the first structures in the universe came into being. But many scientists remain sceptical. 'It's hard to say,' replies Volonteri diplomatically when I ask if primordial black holes could be the true origin of supermassive black holes.

SUPERMASSIVE COLLISIONS

As Volonteri and I are talking in the gardens of the Paris Observatory, the Milky Way and the Andromeda galaxy are travelling towards each other on a collision course. 'In the universe, everything moves,' says Volonteri, 'nothing is static, so eventually different objects can collide with each other.'

The majestic galaxies that look so peaceful on astronomers' images are morphing over epochs of a scale human consciousness can hardly conceive of. The Milky Way and the Andromeda galaxy will collide in roughly four billion years.[12] It certainly gives a new perspective on our human affairs: while we here on Earth are going about our daily business, we're also in the preliminary stages of a head-on galactic collision.

When two galaxies merge it's not like two cars colliding. Galaxies consist primarily of empty space; the distances between the stars are vast. A galactic collision is therefore more reminiscent of two buckets of sand being thrown at each other. Just as the grains of sand are mixed together, so too are the stars and all the gas in the two galaxies, except that in the latter case, the process can take billions of years.[13]

In this hybrid galaxy-to-be, which has been dubbed Milkomeda, Sagittarius A* and the supermassive black hole at the centre of Andromeda might eventually make their way towards each other. 'The expectation is that the two black holes will merge,' says Volonteri. 'If anyone is alive then, they can observe a burst of gravitational waves, but I definitely will not be there.'

As Volonteri tells me about this future merger I imagine the two supermassive black holes circling each other and then joining to form a new, even bigger black hole. Powerful gravitational waves are sent out in all directions. Soon afterwards, an enormous jet is flung out as gas that has been whirling around in the galactic chaos of Milkomeda falls in towards the newly formed black hole. If people have managed to survive until this distant future event, they can expect to experience a cosmic light show unmatched by anything seen before.

Supermassive black hole collisions may have been common in the dawn of time. The universe was denser back then, and galaxies collided more frequently, forming new, larger galaxies. The black holes at their centres could also collide, merging and growing. This gives us another possible explanation for how black holes can grow, as well as ingesting gas or stars. But how often it occurred is unclear. 'Unfortunately, the galactic journey of black holes is complicated,' says Volonteri. 'It can be difficult for two black holes to reach each other.'

To find out how often such collisions happened all that time ago, astronomers must observe the gravitational waves they created. LIGO and the other detectors on Earth are not capable of measuring gravitational waves of this kind, as their frequency is too low. It will, however, be possible to detect them with the aid of the space observatory LISA. 'We will enter a world that cannot be seen in any other way,' says Volonteri. 'It's a totally new perspective on spacetime that opens to us.' Unfortunately, LISA requires major technological advances and will not be deployed before the mid-2030s, but as luck would have it, nature has given astronomers another detector that can register gravitational waves from supermassive black holes: our own galaxy. When these

THE ORIGIN OF THE GIANTS
181

gravitational waves pass the Milky Way, the distance between the stars changes. The effect is extremely tiny, in the order of a ten-thousandth of a millionth of a millionth relative distance shift between each and every star. For instance, the distance between the Sun and our solar system's nearest star, Proxima Centauri, changes by only a few hundred metres. The frequency of the gravitational wave is so low that it can take several decades for this oscillation to shift back and forth.

If it were possible for astronomers to measure these shifts in the distances between stars, they would be able to tell whether the gravitational waves from colliding supermassive black holes pass the Milky Way. Thanks to pulsars – neutron stars that rotate at dizzying speeds and send out regular radio signals – it is. When the distance between Earth and a pulsar changes, it affects the time it takes for the neutron star's radio pulses to reach Earth. By carefully tracking the signals from several pulsars, astronomers can therefore use them like galactic gravitational wave detectors. The method is known as Pulsar Timing Array.

On 29 June 2023, about two years after my meeting with Marta Volonteri, press conferences were held by North American, European, Indian, Chinese and Australian research teams. They were sharing the news that they had succeeded in measuring the passage of gravitational waves through the Milky Way by following a number of pulsars for more than fifteen years. Unlike LIGO, they did not identify individual gravitational waves. Instead, they had observed the effects of all the low-frequency cosmic rumbling created by supermassive black holes rotating around and colliding with each other.

In order to undertake this measurement, astronomers needed to know not only the exact positions of the telescopes on Earth (because the movements of the continents alter the distance between the telescopes and the pulsars), but also the position of

Earth and the solar system in space. When the space probe Juno arrived at Jupiter in 2016 it measured the strength of Jupiter's gravitational field. Thanks to this information, astronomers were able to confirm Earth's position and the midpoint of the solar system (which does not coincide with the midpoint of the Sun) to an accuracy of just a few hundred metres. This was the level of accuracy needed for them to detect gravitational waves with the aid of pulsars.

There are other possible causes for gravitational waves of this kind; aside from the collision of supermassive black holes, they could also have been created by extreme conditions immediately following the Big Bang. Soon, increasingly precise measurements will make it possible to confirm the exact origins of the gravitational waves – and to gather more information on how often supermassive black holes collide.

A LONG, QUIET LIFE

Listening to Marta Volonteri's tale of the mysterious origin of supermassive black holes, I find two contradictory thoughts arising within me. On the one hand, black holes are dark, peculiar objects. They consist of space and time and are difficult to conceptualize on the basis of our experience on Earth. On the other hand, black holes that are created from collapsing clouds of gas and dying stars are formed predominantly from the most common elements in the universe, hydrogen and helium. How can something so simple and ordinary give rise to something so strange and exotic?

'The hydrogen and helium turn into something that we can no longer call hydrogen and helium,' says Volonteri. 'When we study the light of stars, we can infer what they are made of.

THE ORIGIN OF THE GIANTS

But if they turn into black holes, the information about what they were disappears.' All that is left is a dark object in space, a cosmic mystery that astronomers dedicate their lives to solving. Volonteri is laconic as she summarizes the difficulty: 'The origin of a supermassive black hole is not written on its T-shirt.'

Even so, I ask her if she might venture a guess as to the origin and evolution of Sagittarius A*. After all, it's our closest supermassive black hole; surely we want to know where or what it came from. 'From statistical reasoning based on galaxies with similar properties to the Milky Way, I can say what I expect,' Volonteri says. 'Most of the mass of Sagittarius A* comes from the accretion of gas. The black hole has had few mergers with other black holes. Its rotation is about 70 per cent of the maximum possible.'

But exactly what Sagittarius A* was created from – if its seed was a primordial black hole, an imploding star or a cosmic gas cloud – Volonteri is reluctant to say. What she *does* say is that Sagittarius A* has had 'a very quiet life, for a very long time'. Perhaps it was once the engine of a jet-emitting quasar, but today the black hole has such a retiring existence it would be hard for an astronomer in a nearby galaxy even to notice it was there.

Suddenly Volonteri gives a little shriek. A dog races past our legs, chasing a ball. 'I was bitten by a dog recently,' she says, clearly shaken. 'I still have the scars.' Our conversation is drawing to a close, and we get up and walk towards the exit. The dog makes me think of the contrast between the day-to-day life of a scientist and the objects they study. Marta Volonteri's work concerns one of the largest and most terrifying objects in the universe, as she wrestles with theories on how these cosmic monsters came into being at the very dawn of time. And yet it's the dog that alarms her. There are plenty of things to be afraid of on our own planet; black holes in outer space are the least of her concerns.

'To be honest, I'm at a loss,' Volonteri sighs as we pass through the garden's iron gates. 'I've tried everything for the last twenty years, but the deeper I dig, the more problems I find. I'm running out of ideas.' Coming from one of the leading experts on the origin of supermassive black holes, this level of hopelessness is a clear indicator of how difficult a puzzle this is.

We say farewell and Volonteri returns to her office to continue her research. While I continue my stroll along the sunny streets of Paris, I consider the ways people have explored the darkness of black holes. These strange objects found their way into our consciousness in the form of equations; they were mathematical ideas. After extensive analysis, scientists concluded that black holes can form from the death of stars, but it wasn't until after this conclusion had been reached that this kind of black hole was first observed. We knew how they arose before we'd even observed one.

The opposite was true of supermassive black holes. They were observed first, in the form of the vast power of quasars. It was only then that scientists needed to explain where they came from. Even today, half a century after the first observations, their origins are unclear.

But two years after my meeting with Marta Volonteri, a breakthrough takes place.

A NEW WINDOW ON THE UNIVERSE

'Doing research about black holes gives me an adrenaline kick,' says Priyamvada Natarajan (or Priya, as she is often known). She is an eminent professor at Yale University, and is currently

Chair of the Department of Astronomy. I meet her for lunch at an Italian restaurant in Cambridge, Massachusetts, on an early summer's day in 2023.

Priya Natarajan has just given a talk at a conference at the Sheraton Commander Hotel, near Harvard University. The conference has been organized by the Black Hole Initiative, which she is involved in running. Experts from around the world have gathered to listen to the latest results in black hole research.

After the conference we walk through a low-rise residential area of picturesque streets warmed by the May sunshine, until we reach the restaurant. Natarajan slices her pizza as she continues her tale: 'When I was younger I tried several adventure sports like skydiving and other crazy things. I was pushing the limits of my physical ability. With black holes it's the same, I feel like I'm pushing my intellectual limits.'

Like Marta Volonteri, Natarajan is one of the world's foremost experts in the origin and evolution of black holes. 'We are good friends,' Natarajan says of Volonteri. 'We have often worked together.'

I've arranged to meet Natarajan because I want to hear more about a discovery she has recently made that could revolutionize our understanding of how black holes are formed. But I'm also curious about the personal and professional backgrounds of the scientists who have devoted their lives to understanding these strange objects. What path did Natarajan, who was raised in New Delhi, India, take on her way to becoming the Chair of the Department of Astronomy at one of the world's most prestigious universities?

Natarajan was born in Coimbatore in southern India, but her family moved to Delhi shortly afterwards. 'I was five years old when my parents bought me a telescope,' she says. 'We lived in a beautiful part of the city. There wasn't much light pollution and

I could see the stars almost every night. I was an avid amateur astronomer and I fondly remember my dad driving me to the outskirts of the city to see Halley's comet through the telescope. It was incredible.'

In the late eighties Natarajan won a scholarship to travel to the US and study at MIT. She met students from all over the world and devoured courses on every possible subject. Through a research project, she met Martin Schwarzschild of Princeton University. He had followed in his father Karl Schwarzschild's footsteps and become a successful astrophysicist. 'Martin was old and in poor health,' Natarajan recalls, 'but we got on very well and spent a whole weekend chatting.' Schwarzschild told her that if she did a PhD at the University of Cambridge in the UK, she would be able to develop new models of black holes with some of the foremost black hole scientists, such as Stephen Hawking and Martin Rees.

Natarajan followed Schwarzschild's recommendation. After submitting applications and references she was accepted as a doctoral student under Martin Rees at Cambridge's Institute of Astronomy. When she got there she was stunned by the rural setting. 'MIT was in the centre of the city,' she tells me, 'while the Institute of Astronomy was on the outskirts of Cambridge. In front of the institute was a meadow where cows grazed. But it was an amazing environment with incredible people.'

In the late nineties, as Natarajan was doing her PhD, the astronomer John Magorrian and his colleagues made an important discovery. They had studied thirty-six galaxies and their supermassive black holes using the Hubble Space Telescope. Magorrian had identified an unexpected correlation: the greater the mass concentrated in the central part of the galaxy, the greater the mass of the black hole at its midpoint. Around the same time, two groups of astronomers showed that the greater a

THE ORIGIN OF THE GIANTS

supermassive black hole's mass, the faster the stars in that galaxy travelled. The reason these correlations were so surprising was that gigantic black holes, despite their name, are still very small in relation to their host galaxies. Since the black holes cannot have a gravitational influence on the whole galaxy, the link indicated that supermassive black holes share a common evolutionary history with the galaxies they reside in.

'We started to realize that black holes were an important part of the overall cosmic ecosystem,' recalls Natarajan. 'There was a connection between the stars that formed in a galaxy and the mass of its central black hole. The black holes had to play an important role in shaping the galaxies because they had ensconced themselves at the centre of most, and perhaps all, galaxies. But there was no origin story.'

Natarajan began to investigate how these gigantic black holes had developed in relation to dark matter, galaxies and the universe as a whole. She developed a model in which all the radiation generated when matter falls into a black hole influenced the characteristics of the galaxy as a whole. 'Black holes were seen as objects that gobble matter, but at the time it was not recognized that they could be engines ejecting energy, driving phenomena on a larger cosmic scale,' Natarajan says. 'My papers explaining all these new observations unfolding at the time were seen as radical because the ideas proposed and explored were speculative. It took a while for people to get convinced.'

In 2006 she and the astrophysicist Giuseppe Lodato launched a detailed model of how gigantic black holes could have been formed at the dawn of the universe.[14] They suggested that a gas cloud in close proximity to a group of giant stars sending out powerful ultraviolet radiation could collapse directly into a black hole. They showed how such direct collapse black holes formed and evolved in relation to the presence of both dark and ordinary matter, and that the black holes could have a mass roughly equal to the mass of the galaxy they ended up in. This relationship differs from the way things are today, when the mass of a supermassive black hole is less than 1 per cent of the mass of their host galaxy. In the case of the Milky Way, the mass of Sagittarius A* is even less: in the order of a ten-thousandth of a per cent of the mass of the galaxy. But according to the model Natarajan was involved in developing, this relationship could have been very different in the universe's infancy. Together with members of her research group, she predicted that this type of black hole would be possible to see with the new James Webb Space Telescope (JWST).[15]

A RECORD-BREAKING TELESCOPE

On Christmas Day 2021 the JWST was launched on board an Ariane 5 rocket from the Guiana Space Centre in Kourou, French Guiana. It had taken more than twenty-five years for NASA, the European Space Agency (ESA) and the Canadian Space Agency to construct the telescope. The price tag was $10 billion. 'The whole project was at the point of being cancelled multiple times because it was over-budget and delayed,' recalls Natarajan. 'We didn't know if the Webb Telescope would launch.'

After a journey of 1.5 million kilometres the telescope reached its observation point beyond the Moon's orbit. It unfurled its gold-plated mirrors and began collecting the infrared light it was specially designed to pick up.

The telescope worked perfectly. Its first images of dust clouds and galaxies captivated astronomers and the public alike. One of the pictures showed thousands of galaxies shining in different colours against the darkness of space. The light from some of

THE ORIGIN OF THE GIANTS   **189**

them had been travelling for more than thirteen billion years. 'The telescope opened a new window into the early age of the universe,' says Natarajan. 'Before that, everything was an extrapolation from things we'd observed closer to us in the universe.'

Astronomers were surprised by the telescope's first observations. It seemed that, in the universe's infancy, many galaxies had grown incredibly rapidly. They shone brightly and contained a lot of gas and stars. Some of them looked as evolved as our own Milky Way, but it wasn't just the size of the galaxies that astonished astronomers. It also seemed that, in those early days of the cosmos, supermassive black holes abounded and thrived.[16] New black hole records were continuously being set by the JWST. In January 2024 astronomers reported that they had identified the oldest-known black hole in the galaxy GN-z11.[17] The light from the galaxy was emitted 400 million years after the Big Bang (when the universe was less than 3 per cent of its current age) and had previously been observed by the Hubble Space Telescope. But thanks to the JWST, astronomers were able to identify its central black hole and deduce that it had a mass equivalent to around 1.6 million Suns. A few months later, JWST managed to spot the earliest mergers of two supermassive black holes, which happened some 740 million years after the Big Bang, and towards the end of 2024 another record was broken as a black hole in the early universe was caught feeding at one of the fastest rates ever observed.[18]

But the talk of the town among astronomers was a peculiar type of galaxy observed by the JWST. The telescope spotted several bright galaxies glowing in the infrared part of the spectrum. These galaxies were surprisingly compact – only a small percentage of the size of the Milky Way – and were quickly dubbed 'little red dots'. They appear to have been common during the first billion years after the Big Bang, after which they are no longer spotted. The nature of these galaxies remains a topic of intense debate among astronomers. The unusual properties of their light could indicate an extremely dense region of stars, but another intriguing possibility is – you guessed it! – burgeoning supermassive black holes. In this case, the observed light would come from gas clouds accelerated to extreme velocities by the black holes. There are various factors that can contribute to the red hue of the galaxies, ranging from the way light has been redshifted as it has traversed the universe (more about that in Chapter 14) to dust obscuring the black holes. What has truly puzzled astronomers is that the mass of these black holes (if that's what they are) appears to be comparable to the total mass of their host galaxies – a discovery that lends credence to the idea that the black holes were formed by direct collapse. These 'little red dots' may hold crucial clues to how structures formed in the early universe, but their true nature remains an open question.

For Natarajan, however, the peculiar and surprising galaxy UHZ-1 proved particularly intriguing. 'I almost fell off my chair,' Natarajan says, recalling her reaction when she saw the light the telescope had collected from UHZ-1. A group of astronomers had shown that this red dot was a galaxy whose light had been travelling for over 13.2 billion years.[19] 'It feels nearly miraculous that we can receive photons from a time when the universe was merely 3 per cent of its current age,' Natarajan says.

The James Webb Space Telescope shouldn't actually be able to see such dim objects from so early in the history of the universe, but nature has rewarded astronomers with a wonderful tool that allows them to study these distant objects. When light passes a gigantic cluster of galaxies, it is bent and focused by the gravity of their collected matter. The cluster acts as a cosmic magnifying glass that makes even fainter, more distant galaxies visible. In the case of UHZ-1 it was Abell 2744, also known as Pandora's Cluster, that served as the magnifying glass. To Natarajan, Pandora's Cluster was like an old friend: she had been part of a team that produced a detailed map of its distribution of dark matter, thereby calibrating the power of this celestial lens.

In order to find out what kind of black hole UHZ-1 contained, Natarajan and her colleagues needed additional data that required a further telescope. A short walk from the restaurant where we have lunch lies the Center for Astrophysics | Harvard & Smithsonian, where the Hungarian astrophysicist Ákos Bogdán works.[20] He is an expert in the analysis of X-rays, and together with Natarajan and several other collaborators, he used the space telescope Chandra X-ray Observatory (named after the Nobel Prize laureate Subrahmanyan Chandrasekhar) to explore the region inhabited by UHZ-1.

Bogdán and his colleagues saw several X-ray photons emanating from the galaxy. 'X-ray photons are a clear-cut signal of a growing black hole,' Natarajan explains. The high-energy photons are created by matter moving at dizzying speeds around the gigantic black hole in the middle of the galaxy. When Natarajan and her colleagues analysed the infrared light the JWST had observed alongside the X-ray photons, they were able to confirm that UHZ-1 contained a black hole of up to a hundred million solar masses.[21] This was roughly as massive as all the stars in the galaxy – which was exactly the kind of relationship between a galaxy and its supermassive black hole previously suggested by Natarajan and her colleagues. 'UHZ-1 gives strong evidence that black holes can form via direct collapse,' she says.

As I'm listening to Natarajan I try to imagine how this happened. A hundred million or so years after the Big Bang the universe was mostly a void. Space was full of dark matter and scattered dust clouds that consisted mainly of hydrogen and helium. A small number of giant stars burned brightly, solitary in the darkness. In this cosmic dawn, a vast gas cloud gradually began to collapse. Showered with the ultraviolet radiation from a nearby star cluster, a darkness began to grow from within the imploding cloud: a gigantic black hole. It grew in size as more and more gas fell into it. All the while, it was travelling towards the star cluster and together with these stars it evolved to form a larger galaxy. A considerable time after that some of its light began its long journey through the universe until it at last reached our solar system and the James Webb Space Telescope.

'The discovery of an object like this is quite transformative,' says Natarajan. 'We have solved one of the most important problems, which is whether there are more ways for black holes to form than from dying stars. The answer is yes.'

COSMIC DETECTIVE WORK

Marta Volonteri was involved in the discovery of UHZ-1 too. 'I was excited,' she says when I meet her again sometime after my lunch with Priya Natarajan. 'This shows a possible way to form black holes. It fitted very well with Priya's idea.' But Volonteri emphasizes the fact that the JWST and the Chandra X-ray Observatory can see UHZ-1 because it is such an extreme object, which means its supermassive black hole is not representative of all black holes of its age.

Over the coming years astronomers will study UHZ-1 in more detail.[22] They must rule out other explanations for the galaxy's radiation. Only then can they be sure it is a supermassive black hole that was created in the way Natarajan and her colleagues suggest. 'Every new discovery leads to hundreds more questions,' Natarajan reflects. 'I enjoy the fact that we don't see an end to our explorations. The discovery of an extreme object like

THE ORIGIN OF THE GIANTS

UHZ-1 answers a difficult question about the origin of black holes, but it also opens the door to many new questions, like how often black holes are created from direct collapse instead of when stars die.'

In the 2030s several new observatories, such as the Extremely Large Telescope in the Atacama desert in Chile and the gravitational wave detector LISA, will be able to see a long way back through the history of the universe. At that point, astronomers will probably be able to determine, once and for all, all the ways in which supermassive black holes can form.

But even if the mystery of the origin of the black holes is solved, that doesn't mean we will fully comprehend these objects or the wider universe. 'I don't think we will ever be able to say that we understand the entire cosmos,' said Marta Volonteri during our conversation in the garden of the Paris Observatory. 'There is always something new that we need to explain.'

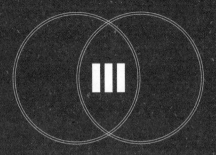

BLACK HOLES AND OUR PLACE ON EARTH

A cosmic darkness so big, so empty, so terrible in its absence of content and meaning that no being, no thought, nothing, can survive there. That is the ultimate future of the universe, so far off in time that the numbers used to describe it get lost as they add up and up.

There are few survivors. The only ones left now are those who appeared in space at the very beginning, those who saw the birth and the end days of galaxies, who saw stars extinguished and species go extinct. The supermassive black holes have lived on as a darkness within the darkness.

But they too will eventually disappear. Slowly they will dissipate in the vacuum that surrounds them. A faint glow, a final electromagnetic sigh – and then only one thing will remain: the ever-expanding void.

Chapter 10

. . . .

PŌWEHI AND THE
RIGHT TO THE LAND

'That day turned my world upside down.'
Jessica Dempsey takes a deep breath and begins to recount a series of events that took place several years ago. I have come to meet her in her office at ASTRON, the Netherlands Institute for Radio Astronomy, where Dempsey is the director. She is also a member of the Event Horizon Telescope organization. On the wall behind her desk is a painting of the Pleiades star cluster by an Aboriginal artist from Australia, Dempsey's country of origin.

The day in question was 2 April 2015. The phone rang in the house in Hawai'i where she was living. At the time, she was working at the James Clerk Maxwell Telescope on Maunakea. One of Jessica Dempsey's colleagues informed her that the astronomers who'd spent the night at the telescope were unable to get down the mountain. The road had been blocked by representatives from Indigenous groups. She rushed out to her Jeep and drove to Maunakea, concerned for the safety of her staff. 'Working at that altitude can be very dangerous,' she says.

Dempsey was accustomed to working in inaccessible places. She'd grown up on a cattle farm in the Australian desert. After studying astrophysics in Sydney she was given the opportunity to travel to the Antarctic. At the age of just twenty she became the first Australian woman to set foot on the South Pole. Her job was to test equipment for a radio telescope. She returned to the South Pole many times and eventually lived through the dark half-year of the Antarctic winter. By the time her stint was over she had permanent frostbite damage on three fingers and felt uneasy in large groups. 'I worked alone and became an introvert,' she says.

Jessica Dempsey struck out for warmer climes. She started working at the James Clerk Maxwell Telescope thinking life in Hawai'i would be mostly sandy beaches and tropical heat. But it's cold at the top of Maunakea, and Dempsey's first task was to shovel snow. 'I should have learned Hawaiian before I went there,' she jokes, 'because Maunakea means the "white mountain".'

Dempsey worked her way up through the ranks and was put in charge of the day-to-day running of the telescope. That's why it was her responsibility to get the astronomers who were trapped on the mountain back down. The blockade was a response to the start of construction of one of the world's biggest telescopes. State and private funders from the US, Canada, China, Japan and India were footing the bill – a cool $1.4 billion. The telescope was called the Thirty Meter Telescope (TMT). It was going to make it possible to study everything from the first galaxies in the universe to planets orbiting stars in the Milky Way, all with a level of detail that had never previously been possible. But Native Hawaiians had another name for it: Too Many Telescopes.

Many Native Hawaiians were critical of the telescopes on Maunakea. In 1964 the Dutch astronomer Gerard Kuiper of the University of Arizona travelled to the mountain.[1] On behalf of NASA and the university, he investigated whether the site was suitable for astronomy and installed a telescope. The high altitude and perfect weather conditions made the mountain one of the world's best locations in which to conduct astronomy. Fifty years

PŌWEHI AND THE RIGHT TO THE LAND   **199**

later the mountain was home to thirteen observatories, many of which have been used by astronomers to learn more about black holes. The most famous example is the Keck telescopes used by Andrea Ghez and her team to map the movements of the stars around Sagittarius A*.

But the critics objected to the astronomical facilities on the basis that they damaged the natural environment of Maunakea, that they were visible from a great distance, spoiling the view of the mountain, and, perhaps most significantly, that they destroyed its sacred status. Maunakea plays a central spiritual and cultural role. It has also been referred to as Mauna a Wākea, which means 'Mountain of Wākea'.[2] In Hawaiian mythology, Wākea is the male god of the sky. Legend has it that the Hawaiian islands were created by Wākea and the goddess of the Earth, Papahānaumoku. The snow goddess Poli'ahu lives on the mountain. The privilege of scaling the mountain's peak had previously been one reserved for Hawaiian leaders.[3]

The planned construction of yet another telescope was the last straw for many Native Hawaiians. 'My ancestors travelled across the Hawaiian Islands by following the stars,' singer Hāwane Rios told the media. 'So while astronomy is an important part of our culture, it is completely absurd to us that the TMT project will have negative impacts on the environment. To see this construction on the summit is just heartbreaking.'[4]

And so, when the construction vehicles headed up the mountain to start work on the telescope, they found a large group of people blocking the road.[5] These people felt that for decades the mountain had been treated poorly by astronomers and by the University of Hawai'i, which assigns land rights for the telescopes. They felt they had been left out of the decision to build another telescope, and wanted to protect Maunakea from further exploitation. They called themselves *kia'i*, guardians.

Dempsey was forced to stop on the road halfway up the mountain. There were rocks on the tarmac and more than a hundred people had gathered around them to prevent the excavators and trucks progressing any further. Dempsey saw a sign that read: *''a'ole TMT'* – 'No to TMT'.

'I rolled down the car window to talk to them,' she says. 'But they only wanted to talk to someone who knew Hawaiian. I felt humiliated. Not because they stopped me, but because it was terrible that I couldn't answer in Hawaiian. It was arrogant to move to an island and not even bother to learn the language that people on the island speak.'

Dempsey explained in English that she was a representative of one of the telescopes and that the telescope's staff needed to come down the mountain. 'They were very polite,' she says of the people blocking the road. 'They understood that I was concerned about the safety of the astronomers.'

The guardians of Maunakea moved the rocks so that she could drive up the mountain and fetch her staff. As she was driving back down, she contemplated the experience. Dempsey was an astronomer. She looked after a telescope that taught people more about the universe. She'd never thought anyone could be against that. 'I felt that I was on the wrong side,' she says. 'I didn't want to be a bad person. But we astronomers were very disconnected from reality. It's all about privilege. The Indigenous people have the lowest incomes, the worst education, and any claim that astronomy creates jobs for them is just bullshit.'

After the blockade, work on the construction of TMT was temporarily halted. Several of the guardians of Maunakea were arrested by the police. Dempsey saw it as her responsibility to set up meetings so that the astronomers could get to know the people involved in the blockade and understand why they wanted to prevent the new telescope being built. At one meeting a participant complained that, despite having lived on the islands for many years, astronomers had never mixed with Native Hawaiians, that they themselves weren't allowed to visit the telescopes, and that their children couldn't get jobs there.

'It hit me hard,' Dempsey recalls. 'We had done everything wrong. We were not prepared. We had no ethical foundation. And why was that? Because of the attitude that the acolytes of Western science are taught: we don't need ethics because we are objective.'

Around the same time, the astronomer Sandra Faber emailed her colleagues.[6] She's a professor at the University of California Santa Cruz, and belonged to a research group that had used the Hubble Space Telescope to show that there were enormous black holes at the centre of many galaxies. In her email, Faber encouraged her colleagues to sign a petition in support of TMT's construction. Many were shocked at her choice of words when she wrote that the telescope had been 'attacked by a horde of native Hawaiians who are lying about the impact of the project on [the] mountain and who are threatening the saftey [sic] of TMT personnel'.

Faber's statement was roundly criticized. 'She felt entitled to make such comments,' Jessica Dempsey says. 'It shows what disconnected bubbles we have created in our scientific communities. Many astronomers were horrified and ashamed.' Faber swiftly apologized, but for Dempsey the email reinforced the sense of disparity between the astronomers and parts of the population on Hawai'i.[7] 'We often joke about the academy's ivory tower,' she continues. 'But that is exactly what we have created. Many scientific institutes and universities pride themselves on how exclusive they are and how difficult they are to access. We have created a language that is inaccessible to people and then we wonder why we have a credibility problem.'

Conflicts around the planned construction of TMT continued in subsequent years, both through protests and legal appeals.[8] Dempsey set about trying to understand the root of the problem. She started studying intersectionality theory, which describes how socioeconomic class, ethnicity, gender and other identities combine to influence relationships of power. She looked into the colonial history of the Hawaiian islands and began to question her view of astronomy's role in society. 'You may have worked hard to get rid of your own prejudices and do good things in life,' she says, 'but you can still represent colonial oppression. The two things do not cancel each other out. Being able to ground yourself in those two opposing truths is hard, but it's very important.'

Jessica Dempsey realized that if astronomers were to be able to continue with their work in an ethical way, they had to face up to their own colonial history.

COLONIAL ASTRONOMY

On 18 January 1778 Captain Cook arrived at the Hawaiian islands with two British Navy ships. He had travelled from Great Britain, passing the cape of South Africa and travelling on past Australia and New Zealand towards the vast Pacific Ocean, where he eventually became the first European to set foot on Hawai'i.

Just like the Polynesian wayfinders who had arrived at the volcanic islands more than a thousand years previously, Cook navigated the waves using the stars. By studying their position in the sky he was able to determine the ship's latitude – that is, what their position was along a north–south axis. But finding out the longitude – the east–west position on the globe – was a problem the British Navy had long sought to solve. This knowledge was crucial to the European nations sailing across oceans

PŌWEHI AND THE RIGHT TO THE LAND

to expand their empires, as royal houses and trading companies raced to lay claim to 'new' lands.

The British government had promised a large sum of money to anyone who could find a reliable method for determining longitude at sea. John Michell – the man who initially proposed the idea of dark celestial bodies – was a member of the Board of Longitudes, which judged the proposals. One of the main techniques used a clock set to the time at Greenwich, England. A navigator at sea would measure the position of the Sun to deduce the time of local noon (when the Sun is at its highest point in the sky), and then compare that time with the Greenwich time on the clock. From this time difference, the ship's longitude could be calculated, but it required the development of precision chronometers that could keep time accurately, regardless of humidity, temperature variations and the swell of the sea. It was a clock approved by Michell and his colleagues that Cook had with him when he reached the Hawaiian islands.[9] Astronomical observations and expeditions, coupled with the development of better clocks, were thus an integral part of the expansion of empires and the scientific mapping of the Earth's surface.

According to some scholars, Cook was originally hailed as a god by the people of Hawai'i. This divine aura soon evaporated, however, with the violence, brutality and theft that followed in Cook's wake. One example of his many transgressions was his unilateral decision to take the wood from a burial site when he needed to repair one of his ships. 'The poor dismayed chiefs,' wrote one of his crewmen, 'behold the fence that enclosed the mansions of their noble ancestors, and the images of their gods torn to pieces by a handful of rude strangers.'[10] The conflicts grew, and Cook eventually met his fate on the beach at Kealakekua. He had tried to kidnap King Kalani'ōpu'u, but the islanders resisted, overpowering Cook and several of his crew. Cook was stabbed to

FACING INFINITY

death with a knife the islanders had previously obtained from him in a bartering exchange. The rest of Cook's crew fired cannons at the beach and several of the islanders were killed. Thus ended the first encounter between Europeans and the people of Hawai'i.

Not that Cook's death hindered other Europeans returning to the islands. King Kamehameha I united the islands into one royal kingdom in 1810. It was recognized early as an independent state by the US and European powers. But in 1820 Christian missionaries began spreading their religion on the islands, and a few years later the first sugar plantations, which would fundamentally transform the Hawaiian economy, were established. The Indigenous population lost control over several areas of land, and many died of illnesses brought by European immigrants.[11] It is estimated that when Cook arrived, the population of the islands was 400,000. A hundred years later it was only 40,000. Many Japanese, Chinese, Filipino and Portuguese families migrated to the islands to make up for labour shortfalls on the plantations and the businesses that sprang up around them.

As the economy and the plantations grew, conflicts began to escalate over land rights on the islands. Foreign landowners with links to the US wanted increased control over the fertile soil. In 1887, backed by a 500-strong militia, they forced the king to put his name to a new constitution that radically decreased his power as well as the rights and influence of the Native people.[12] At that time, Hawai'i was an impressive nation state. Its population boasted one of the world's highest literacy rates. The king also had electricity and a telephone installed in his newly built palace long before either the White House or Buckingham Palace could claim such luxuries.[13]

But conditions for the local populace were about to deteriorate dramatically. When Queen Lili'uokalani tried to overturn the new constitution, the white landowners refused to agree.

With the aid of the US Marine Corps they took control of the government buildings in 1893, imprisoning Lili'uokalani in her own palace and forcing her to abdicate. Five years later, the US government annexed Hawai'i, and the islands' fertile lands were brought under US administration. A coup d'état had taken place, and the Hawaiian Kingdom was overthrown.[14]

The events of the late nineteenth century still affect the situation on the islands. In 1959 Hawai'i became a US state and administration of the land was transferred to the state government. Nine years later, the University of Hawai'i was given the task of allocating land rights for new telescopes on Maunakea. The lease held by the university was to last sixty-five years.

The telescopes on Maunakea are not the only astronomy project to bear the stamp of colonialism. Take the most famous expedition to observe a solar eclipse: the British astronomer Arthur Eddington's observations on the island of Príncipe off the coast of Gabon in West Africa. As we saw in Chapter 8, he travelled there with his colleagues in 1919 to observe a solar eclipse, intending to test a prediction of Einstein's general theory of relativity. Príncipe was a Portuguese colony at that time. Only a few decades before Eddington set foot on the island, an estimated 67,000 Africans, mostly from Angola, had been forcibly relocated there to work on the island's cocoa and coffee plantations.[15] Although slavery had officially been abolished in the late nineteenth century, the colonial structures lived on: the island was governed by a small number of Portuguese families while the 'freed' slaves continued to work on the plantations. When Eddington arrived, the plantation workers were put to work carrying his equipment through the near-impenetrable terrain to a remote hill where they built a makeshift observatory.[16] Thanks to the observations Eddington and his colleagues carried out, Einstein's theory was confirmed. Both Einstein and Eddington became widely known as a result of these observations, while the workers who carried the equipment remained in obscurity.[17]

Colonial themes are once again associated with Eddington's name around the start of the thirties. The Indian physicist Subrahmanyan Chandrasekhar had travelled from India to the UK to do a PhD at the University of Cambridge. On the boat that carried him over the Arabian Sea, across the Bay of Aden and into the Mediterranean, he had calculated that there was an upper limit to the mass at which white dwarfs can remain stable. This was the first indication that stellar material could collapse indefinitely to form a black hole. But when Chandrasekhar came to present his calculations at a meeting of the Royal Astronomical Society, he was met with derision by Eddington (who went on to call Chandrasekhar's idea 'stellar buffoonery'). There have been speculations that racism may have played a part in this dismissal. Chandrasekhar experienced racism both on his journey to Cambridge and during his time at the university.[18] Despite his fundamental contributions to astrophysics he was never offered a position in the UK, complaining that 'there is some prejudice in giving Indians a definite appointment'. Instead, he had to seek out academic opportunities in the US, where he was forced to start on a lower rung of the academic hierarchy than his white peers. His story demonstrates how colonial, racist thinking was at play even as the earliest calculations of stellar collapse were being conducted, and how these attitudes held back the development of both astrophysics and black hole research.

Even today, colonial mindsets influence decisions about telescopes and space bases, as seen in the following three examples that have particular relevance for black hole research.

The Greenland Telescope at Thule Air Base

In 1953 Danish authorities arrived at the village of Dundas on the west coast of Greenland.[19] They told the villagers, members of the Inughuit group of Inuit Indigenous people, that they had to move. The village was to be the site for a new military base the US Air Force was building to expand the nearby Thule Air Base at Pituffik. The villagers were given only a few days to pack their things, then all twenty-six families were forced to leave their homes, their school and their graveyard.[20] Their homes were later burned to the ground. The hundred or so forced evacuees travelled with dogsleds to Qaanaaq, around 130 kilometres away. There they lived for some months in makeshift tents, before settling in newly constructed wooden houses.

The US Air Force uses Thule Air Base to monitor for potential missile attacks, among other things. But the base is also used for astronomy. The Greenland Telescope became operational towards the end of 2017, and is one of the radio telescopes in the Event Horizon Telescope network. Astronomers are transported to the telescope by the Air Force, but the families that were forced out of Dundas and their descendants are still living in the camp at Qaanaaq. Despite taking their legal battle all the way to the European Court of Human Rights, they still haven't received permission to return. In 2023 the US Air Force changed the name of Thule Air Base to Pituffik Space Base to 'celebrate and acknowledge the rich cultural heritage of Greenland and its people'. In 2025 US interest in the region took a sharp turn, with the newly elected president Donald Trump stating that he sought US control of Greenland, not ruling out the possibility of achieving that goal with military force.

The Esrange Space Center in northern Sweden

The Swedish space base Esrange lies a little over forty kilometres east of Kiruna in Sweden's northernmost county. It's a launch site for rockets and, in the near future, satellites, used for a range of purposes including weightlessness experiments and atmospheric studies. The base can also be used to launch balloon-borne telescopes such as XL-Calibur, which, among other things, observes X-rays from the black hole Cygnus X-1.[21]

Esrange is built on land that has a colonial history. The Sámi Indigenous people inhabit a region called Sápmi, which spans a large area of northern Norway, Sweden, Finland and the Kola Peninsula of Russia. Sweden's colonization of Sámi regions has led to Sámi people losing control of land and resources and being subjected to race biology research, heavy-handed forms of Christian conversion and disrespectful education and language politics. Mines, wind farms, hydroelectric projects and forestry have limited the ability of the Sámi population to live as they choose. Unlike countries such as Norway and Denmark, Sweden has not ratified ILO 169, a UN convention protecting the rights of Indigenous groups.[22]

Esrange was constructed in the early sixties by the European Space Research Organisation (ESRO). The space centre was a European project in which Sweden was a key partner. Because the rockets, once launched, would come down to Earth again, the base had to be surrounded by large areas of sparsely populated land. Four Sámi villages and their reindeer grazing grounds were within the risk zone. The Sámi were invited to collaborate, not wholly without friction, and were offered financial compensation and the provision of protective shelters.[23]

The space centre opened in 1966, with the French general Albert Le Bras as director.[24] He was neither astronomer nor engineer, but a military commander with a background in the French Air Force

PŌWEHI AND THE RIGHT TO THE LAND

and NATO. He had fought an uprising of the Tuareg, a Berber people indigenous to the Sahara, and the ESRO viewed Le Bras's experience in dealing with them as an advantage, since the centre was located in a Sámi area. The Swedish delegation to ESRO was unhappy with Le Bras's appointment, however, and saw to it that he was replaced with a Swedish director.

These events demonstrate the colonial dimension that accompanied the establishment of the Swedish space centre.[25] The fact that the Sámi were forced to adapt to the needs of the base can be seen as one step in a long history of retreat from their way of life. The conflicts are ongoing. In 2019 a court decision was needed to resolve a dispute between Esrange and the Sámi village of Talma, concerning plans to expand rocket operations.[26] A year later a confidential agreement was reached, but the grievances remained.[27] 'There is a bit of condescension towards us on their part,' says Ol-Johán Sikku from one of the nearby Sámi villages. 'They don't tell us everything.'[28]

James Webb Space Telescope

As we saw in Chapter 9, this wildly successful space telescope has given scientists new insights into the origin and evolution of supermassive black holes through the history of the universe. The telescope was launched from ESA's Guiana Space Centre near Kourou in French Guiana.

The French national space agency is in charge of the land the space centre occupies. France's first rocket launch site was in Algeria, a country it occupied from 1830 until the end of the Algerian War of Independence in 1962. After that, France transferred ownership of the space centre to the Algerian government and established a new base in French Guiana.[29] The country had previously been a French colony and was integrated into the Republic of France after the Second World War. Launching rockets from a base in French Guiana has several technical advantages. The proximity to the equator gives the rockets an extra push off the ground as a result of the Earth's rotation. In addition, the remote location and the waters of the Atlantic offer safe crash-down sites, should a launch fail.

The base is guarded by the Third Infantry Regiment of the French Foreign Legion, the same regiment that fought in Algeria during the 1954–62 War of Independence.[30] In 2017 10,000 people demonstrated outside the base. Protests and a general strike had been crippling the country for weeks, and the demonstrators, many of them from anti-colonial groups that included members of Guiana's Indigenous population, demanded improvements to the economic and social opportunities open to them.[31] They complained that money flooded into the base while residents of surrounding areas lived in poverty. 'There are rails to transport satellites onto a rocket, and there's no train, no metro, no night bus,' a protestor told the media.[32]

Riot police stopped the protestors entering the space centre, but a few union leaders and politicians who came in to negotiate with the director occupied an office inside the complex for twenty-four hours. Rocket launches were postponed, and it wasn't until the French government promised emergency investments of up to 2.1 billion euros that the protests were called off and the rocket launches could begin again.[33]

These examples show how the aftermath of colonialism impacts astronomical activities today. Military support for this colonial astronomy extends beyond simply securing and maintaining the sites on which telescopes and space centres stand. Military technology and logistics also play an important role.

PŌWEHI AND THE RIGHT TO THE LAND

The US military fly astronomers to the Amundsen–Scott base at the South Pole where observatories including the South Pole Telescope (part of the Event Horizon Telescope) are situated. One of the largest arms manufacturers in the US, Northrop Grumman, led the construction of the James Webb Space Telescope.[34] Its predecessor, the Hubble Space Telescope, was constructed by Lockheed Martin, another major US corporation that produces military plane and weapons systems.[35] The design of the telescope was based on a US spy satellite.[36] The adaptive optics that enable the Keck Telescopes to peer into the centre of the Milky Way and study its supermassive black hole were originally developed by the Pentagon to surveil Soviet spy satellites. The link between telescopes and military interests goes back at least as far as Galileo, who presented his newly built telescope to the ruling elites in Venice. Galileo said that the telescope '[allowed] us at sea to discover at a much greater distance than usual the hulls and sails of the enemy... in order to prepare for chase, combat, or flight.'[37] A telescope is as useful for watching one's enemies as it is for stargazing. These are just a few examples of the technological and logistical overlaps between astronomy and the military; a comprehensive list would be much longer.[38]

A striking proportion of black hole researchers have also had military connections. As we've seen, Karl Schwarzschild was a lieutenant during the First World War, and spent most of his time calculating the course of ballistic projectiles while he was working on his formula for black holes. Robert Oppenheimer in the US and Yakov Zeldovich in the Soviet Union both worked on nuclear weapons for the military but also made major contributions to our understanding of black holes. When Roy Kerr discovered the formula for rotating black holes, he was conducting research at the United States' Wright-Patterson Air Force Base.[39]

FACING INFINITY

Behind the skygazing astronomer stands the armed soldier. The flow of people, knowledge and technology between the military and astronomy communities could be described as a military–astronomical complex. This complex has increased our knowledge of space and black holes, but it also has a colonial side, with negative consequences for the environment and Indigenous groups (as well as other local populations). People often say that when astronomers investigate space, they are teaching us more about our origins and our place in the universe. But the long-term impact of colonialism means that this knowledge has been acquired at the cost of certain people's exclusion from their original place on this planet.

THE BLACK HOLE OF CALCUTTA

Even the term 'black hole' has a bloody colonial history. On 20 June 1756 Siraj-ud-Daulah, the Nawab of Bengal, attacked Fort William, a British fortification established to protect East India Company trade in Calcutta (now known as Kolkata).[40] Ud-Daulah's intention was to stop the Company expanding. He chased away many of the soldiers quartered there and locked those who didn't manage to flee in a prison cell in the fort, known as 'the black hole'. The British were prisoners in their own cell, and spent the night trampling each other and gasping for non-existent air and water. When ud-Daulah's soldiers opened the cell door in the morning they found more corpses than survivors.

The British prisoners' deaths led to an outcry in their home country. British forces hit back hard at ud-Daulah's troops and executed the Nawab. The event kick-started their colonial takeover of the whole Indian subcontinent. The expression 'the black hole

PŌWEHI AND THE RIGHT TO THE LAND   **213**

of Calcutta' was used widely in British propaganda campaigns as they regained control of Fort William, and it is apparently still in use: the Cambridge Dictionary defines the phrase 'the black hole of Calcutta' as 'an unpleasantly full and hot room'.

A little more than two centuries after the fateful events in Calcutta, black holes turned up again. This time the subject was not a colonial prison, but a cosmic one. In his autobiography, John Archibald Wheeler recounts how he discussed the potential end stages of a star at a conference in the late sixties. After he had used the unwieldy phrase 'gravitationally collapsed objects' several times, someone in the audience called out, 'How about black hole?'[41]

This someone was later identified as the astrophysicist Robert Dicke of Princeton University, who'd brought the expression from his own home. When his kids couldn't find some object they were looking for, Dicke replied it 'must have been sucked into the black hole of Calcutta!'

Wheeler liked the term. He wasn't the first to use it (it had appeared in US journals as early as 1964), but it was thanks to him that its use spread. Wheeler nurtured a whole generation of physicists, several of whom were awarded the Nobel Prize. He is a central figure in black hole research. When he started using the term at conferences and in articles, he did so as if it were the most natural thing in the world, but not everyone appreciated it. In Russia and France, the words for 'black hole' refer to something quite different from a dead star in outer space, namely, an orifice much closer to home.[42] In the Soviet Union, physicists and astronomers used the term 'frozen star' (because, to the outside observer, the collapse appears frozen in time) or 'dark star', while in continental Europe, the term 'collapsed star' was more common. Today, however, the term 'black hole' is well established and is the default in both the scientific literature and popular culture. The term's colonial origins, on the other hand, have been generally forgotten.

Wheeler personifies the link between black holes, colonialism and the military. During the Second World War he contributed to the development of Nuclear Reactor B, located a couple of kilometres from the LIGO gravitational wave detector in the Hanford desert. Uranium was turned into plutonium at the reactor, partly for the first test detonation of an atomic bomb in the Nevada desert, and partly for the atomic bomb that was dropped on the Japanese city of Nagasaki on 9 August 1945. (The bomb that was dropped on Hiroshima consisted solely of uranium, much of which had been produced in the Shinkolobwe mine in what is now the Democratic Republic of Congo.)[43] In the Hanford desert, parts of which have been contaminated with radiation, the memory of how Indigenous people were forced out by white colonizers is still fresh.

After the war, Wheeler worked on the development of the hydrogen bomb. In 1952 he watched from a US Navy ship as the nuclear device was tested on the island of Elugelab in the Pacific Ocean in an operation code-named Ivy Mike. It was several hundred times more powerful than the atomic bombs that destroyed Nagasaki and Hiroshima, decimating the million-year-old corals that the island consisted of and leaving nothing but a huge crater. The 150 people who lived on the atoll the island belonged to had been forcibly evacuated when the Atomic Energy Commission started testing nuclear weapons in the Pacific.[44] The same year Wheeler watched this test detonation, he started studying Einstein's general theory of relativity. Thanks to his work on nuclear weapons, Wheeler had the resources and knowledge to solve the mystery of the fate of stars, and show how a black hole could be formed.

THE EMBELLISHED DARK SOURCE

Let's return to Maunakea. The protests against the TMT brought the colonial side of astronomy into the limelight. When the time came for the Event Horizon Telescope to present the first image of M87*, Jessica Dempsey saw a chance to do the right thing. Two observatories on Maunakea had contributed to the new image of M87*: the James Clark Maxwell Telescope, where Dempsey worked, and the Submillimeter Array, with the astronomer Geoff Bower at the helm.

Dempsey and Bower wanted the supermassive black hole M87* to have a name from Hawaiian mythology. The idea of giving celestial objects Hawaiian names came from the entrepreneur John De Fries, who is Native Hawaiian.[45] 'It was an offer and an olive branch,' says Dempsey. There was one person in Hawai'i who was perfect for the job of choosing the name: Larry Kimura. He was known as the 'godfather of the Hawaiian language revival', or, as Dempsey put it: 'They say that, without Larry, the Hawaiian language would have died out.'

I meet Larry Kimura one sunny afternoon in his office at the College of Hawaiian Language, where he's a professor. Before travelling to Hawai'i I thought most people there spoke the language, but that's not the case. According to the organization the Endangered Languages Project, Hawaiian is 'severely endangered', and UNESCO lists Hawaiian as a critically endangered language.[46] A 2015 survey indicated that fewer than 20,000 people on the islands speak Hawaiian and only 2,000 of these speak it as their first language. Kimura has therefore dedicated his life to saving Hawaiian from dying out. I'm curious to find out how a supermassive black hole became the focus of his struggle.

Kimura was born in 1946. His paternal grandparents were Japanese immigrants, while his maternal grandparents were

Native Hawaiians mixed with European ancestry. 'My parents never encouraged us to speak Hawaiian,' Kimura tells me. 'When my Hawaiian grandmother socialized with her Hawaiian friends I could hear them say in Hawaiian, "Oh, our Hawaiian language is disappearing so fast." They were very sad about it.'

Aside from hearing Hawaiian spoken by some of his family members, Kimura encountered the language when he heard and sang Hawaiian songs at school. 'We didn't know what we were singing but we sang it beautifully,' he recalls.

A major factor that led to the endangered status of the Hawaiian language happened after the coup in 1893. Three years after the new government came to power, they passed a law stating that English would be the language of instruction in the education system.[47] The consequence of this was that each subsequent generation of islanders spoke less Hawaiian. English played a major part in the colonial conquest of the islands, but in the sixties and seventies a movement emerged that questioned the political, social and cultural status quo. Many asked why they weren't being taught Hawaiian, linking this question to one about why Native Hawaiians were treated so poorly. Kimura was part of this movement, known as the Second Hawaiian Renaissance. The battle to revive traditional Hawaiian expressions of culture and identity led to Hawaiian becoming the official language of the state of Hawai'i in 1978.

Kimura started a Hawaiian-language radio programme and worked with a small group to establish the first Hawaiian-language schools for children. Together with the 'Imiloa Astronomy Center, he launched a project called *A Hua He Inoa* (To Call Forth a Name in Honour of), to revive the Hawaiian tradition of naming celestial objects.[48] His hope is that Hawaiian will once again become a living language in its homeland.

In March 2019 Larry Kimura met a group of people who wanted

PŌWEHI AND THE RIGHT TO THE LAND

to give the black hole M87* a Hawaiian name. Aside from Jessica Dempsey and Geoff Bower, among the group at the meeting were the astronomer Douglas Simons, director of the Canada–France–Hawai'i Telescope, and Larry's niece Leslie Ka'iu Kimura, who is the director of the 'Imiloa Astronomy Center.

'I could tell by their tone of voice that they were excited,' Kimura says. Dempsey explained that the image of a black hole looked like a glowing ring around a dark sphere. She asked whether Larry could think of an appropriate name. After thinking for a moment, Larry said, 'Pōwehi.'

The name means 'the embellished dark source of unending creation', and consists of two parts, *pō* and *wehi*. *Pō* means 'the dark source of unending creation', while *wehi* means 'embellished with honour'. This expression is added to words to denote praise: 'We didn't have diamonds, pearls, crowns or gold,' says Kimura, 'we had words.' Pōwehi is pronounced *poe-veh-hee*, with a long 'oh', emphasis on the 'veh' and a short 'hee'.

The word comes from the Hawaiian creation epic Kumulipo, which tells the story of how the world was created from the darkness. '*Kumu* means source,' says Kimura, 'and *lipo* is a word for darkness. So Kumulipo means "the source of darkness" or "the creation from the deep darkness".' In Kumulipo, the fathomless, powerful darkness *pō* is the source of all life, which starts out in the sea. From a simple coral, more complex life forms evolve, move up onto the land and eventually turn into the 'messy area' (in Kimura's words) of human life. The darkness is ever-present in all these transformations. '*Pō* is the most basic ingredient,' explains Kimura. 'If you don't have it, you can't have life. *Pō* is a powerful darkness that is said to still exist today. It is endless and constantly recreates itself.'

In Kimura's translation of the Kumulipo, the section where Pōwehi first appears reads:

218   FACING INFINITY

> Tumultuous is the fast flow of time
> Slipping into one continuous motion is deep darkness
> It fills, it is overwhelming
> It fills, taking up space, permeating
> Boundless, engulfing, suffusing
> Securing spatial foundations, determining heavenly realms
> It is a celestial flow of time through Kumulipo of deep darkness
> Profound darkness prevails
> Offspring are given birth to pō wehiwehi

Here the darkness has been adorned with a double 'wehiwehi' as a form of embellishment. The name Pōwehi demonstrates the poetic power of the Hawaiian language: a long sentence is condensed into a short word. When Queen Liliʻuokalani was put under house arrest after the 1893 coup she translated Kumulipo into English – the dissemination of the Hawaiian creation story is closely linked with its islands' colonial history.

A NEW NAME FOR AN OLD BLACK HOLE

When Jessica Dempsey heard Kimura's suggestion it sent a shiver down her spine. 'Pōwehi describes in one word what it took us six scientific papers to say,' she says. 'I think it's a wonderful attitude that darkness is not something to be feared, but is actually an engine for change and a place of creation.'

But who has the right to name black holes? The same object in space can have different names in different cultures. In many European languages, for instance, the brightest star in the night sky is known as Sirius, which originates from the Ancient Greek Σείριος (*Seirios* in Latin script), and means 'burning' or 'glowing'. But an old Norse name for the star is Lokabrenna, after the god

PŌWEHI AND THE RIGHT TO THE LAND

Loki, while in China it is known as 天狼星 (Tsien Lang), which means 'the celestial wolf'.

When astronomers discover new space objects without pre-existing names, they have to agree on a common nomenclature. The task of officiating over these names has fallen to the International Astronomical Union (IAU). The organization was founded in 1919 with the aim of promoting international collaboration between astronomers and improving the scientific status of astronomy. The IAU database contains lists of names of newly discovered comets, asteroids, planets, stars, galaxies and galaxy clusters.

However, the IAU does not name black holes. When the organization was founded, no one had ever observed such an object, and so there was no official designation procedure. The names astronomers are in the habit of using, for example Cygnus X-1 or Jo313-1806, demonstrate the imaginative effort astronomers typically expend when it comes to choosing names. The fact that the two best-known supermassive black holes, Sagittarius A* and M87*, contain asterisks that are pronounced 'star' also creates confusion. 'Astronomers shouldn't be allowed to name objects,' Dempsey jokes.

Because the IAU does not name black holes, there was no formal process around the naming of Pōwehi. On 10 April 2019 the image of M87* was released. 'Imiloa Astronomy Center and the East Asian Observatory, which funds the James Clerk Maxwell Telescope, sent out a press release with the news that the black hole had been named Pōwehi.[49] The press release was made available in Hawaiian, the first time an astronomy press release had been translated into the language.[50] 'We saw a huge impact,' she says, 'not just in Hawai'i but all over the world.'

In Hawai'i the name and image of Pōwehi turned up on T-shirts, posters and in the newspapers. The Governor of Hawai'i proclaimed 10 April Pōwehi Day.[51] 'Words can have tremendous power,' Dempsey says. 'Seeing how the name Pōwehi was received in such an open-minded way made me realize how accepting people can be. Being part of the introduction of the name Pōwehi is what I am most proud of. I consider it a scientific achievement.'[52]

The name was well received within the Event Horizon Telescope team too. Just as the darkness plays a creative role in Kumulipo, astronomers have learned that black holes play a creative role in the universe. 'There's a sense of birth to them,' says Shep Doeleman from the Event Horizon Telescope team. 'Black holes shine and radiate as they create jets. So black holes are paradoxical, they are sources of light, they create large structures. The name Pōwehi shows that the Hawaiian Indigenous people realized that darkness can be something that you can revere, it can be something you adorn, it can be the beginning of something new.'

But even though Pōwehi had a major impact, not everyone was happy about the Hawaiian name being spread by astronomers. The student and activist Iwakelii Tong told the media that it felt like an attempt at 'fabricating consent from Native Hawaiians'.[53]

When I ask Kimura what he felt when he saw the image of Pōwehi, he is silent at first. 'It was so beautiful,' he says after a while. 'I've learned that there are millions more black holes. I guess they look the same, but they are bigger or smaller. What is this darkness that has so much power, that is so massive, that there is so much of?'

He removes his glasses and wipes tears from his eyes. 'I'm sorry,' he says. 'I'm not just thinking about the image. I'm thinking about how amazing it is that our people could think about such profound things that people just take for granted.'

I try to understand the emotions Kimura is expressing. They seem to be a mix of pride and distrust – pride in the fact that a word from the Hawaiian creation story has become the name of one of the universe's most powerful objects. And distrust – as I sense it – in the actions and motivations of scientists, who sometimes display a contempt for the experiences of some Native people, and might view them as superstitious and old-fashioned. But who actually has the strongest contact with nature? Those who have risked their own lives when navigating the ocean with the aid of the stars and who are prepared to sacrifice their freedom to defend a mountain? Or the scientists who spend their days behind a computer screen, who process data describing things that happen deep in outer space, but do not notice what is happening to the environment and the people who live around the sites where their research facilities stand?

THE PROTESTS CONTINUE

A few months after the image – and the name – of Pōwehi was presented, thousands of Hawaiians walked up Maunakea. They were once again seeking to prevent the TMT being built. 'I personally am not against science,' Pua Case, one of the movement's leading figures, told the media, 'but I am against science that will destroy our land base.'[54] Case said those protecting Maunakea did so 'in passive resistance, but fierce in the love of our land'.

The guardians set up camp, erecting tents and building bamboo blockades across the road. They spent the night in the cold to protect Maunakea. This blockade too stopped the excavators reaching the top, and construction on the telescope was once again halted. Many people, including several elders, were arrested and the Governor of Hawai'i called a temporary state of emergency. 'The blockade was a masterpiece,' says Jessica Dempsey of the way it was organized. 'There were moments when I wanted to climb over the picket lines.'

It remains to be seen whether the TMT will ever be built. When I visited Hawai'i in the summer of 2023 I got the impression there was a huge distance separating those for and against the construction. Some of those I spoke to were convinced the telescope would be allowed to go ahead. Others were sure that if the astronomers tried to build the TMT, criticism would mount against all the telescopes on Maunakea.

On the road leading to the top of Maunakea, I met Sam Keliiheleua. He was living in a tent with three dogs and was involved in the blockades that had halted the construction of the TMT. 'I'm ready to block the road again,' Keliiheleua said, 'because we don't want any more telescopes built up there. I am defending Maunakea because the mountain is important to my religion, my kids, the Hawaiian language and to learn our culture. It's our sacred mountain.'

In 2033 the land rights for the telescopes on Maunakea will expire, and in the coming years a new administrative body that includes Native Hawaiian representatives will take over management of the telescope land leases.[55] When I asked Keliiheleua what he thought about the new body, he replied, 'I don't know yet. It's an open question.'

The future of the TMT is highly uncertain, not only because of the lack of community support, but also because of funding issues.[56] One thing is for sure, though, and that's that Keliiheleua is prepared to defend the mountain again if necessary.

A NEW KIND OF SCIENCE

After all the work she'd done to introduce the name Pōwehi, Jessica Dempsey was burned out. She relocated from Hawai'i to the Netherlands. 'I told my friends that I'm not the one to

PŌWEHI AND THE RIGHT TO THE LAND 223

end this struggle,' she says. 'But I am deeply grateful for the experience, because it made me not only a better scientist, but also a better person.' For her, the image of the black hole Pōwehi represented a chance for astronomers and local people to form a new relationship. But this relationship remains fragile. When I ask whether she thinks it's possible that all the telescopes will be gone from Maunakea in twenty years, her reply is brief: 'Yes.' She continues, saying, 'I don't think the Western way of doing science works anymore. We need an ethical paradigm shift. The new generation of scientists will be more socially aware and refuse to be involved in projects that are not socially sustainable.'

In her new job as director of ASTRON, the Netherlands Institute for Radio Astronomy, Dempsey has been instrumental in the construction of the Square Kilometre Array (SKA), a large network of radio telescopes that will be built in South Africa and Australia. SKA will, among other things, increase our knowledge of supermassive black holes, but these telescopes too will be built on land belonging to Indigenous people. 'We're building the SKA in one of the most unequal economies in the world,' says Dempsey. She is therefore fighting to ensure that the mistakes that were made on Maunakea are not repeated.

Astronomers have begun to rethink their approach to building new telescopes.[57] They try to make contact with and get the support of local communities before construction begins. Whether this work will lead to more ethically sustainable operations remains to be seen. But thanks to the Native Hawaiian fight to protect Maunakea, more astronomers have been forced to consider the social and political consequences of their research. This issue was brought into sharper focus not least because of a supermassive black hole called Pōwehi.

Chapter 11

. . . .

BLACK HOLES AND
CLIMATE CHANGE

It's the beginning of the twenty-second century. Cities such as Amsterdam, Kolkata, New Orleans, Bangkok and Venice are under water. Most of the island nations of the Pacific Ocean have been drowned. Rising sea levels have made the world we once knew a distant memory. Hundreds of millions of people living in coastal areas have fled their homes.

According to the United Nations' Intergovernmental Panel on Climate Change, the IPCC, this is a possible future scenario.[1] Not only human civilizations are in danger; the risk zones include all life in coastal regions, including whole ecosystems. The main reason is human carbon dioxide emissions. As average temperatures on Earth rise, so do sea levels, but it is hard to predict exactly how much they will rise in a given location. This is because the rise in sea level depends partly on the quantity of carbon dioxide released, and partly on how much the Earth's land masses rise and sink.

To understand which areas risk being submerged, scientists must be able to measure relative changes in the land and sea levels at the level of the centimetre. Only then can they confirm whether the water is rising at a rate that threatens nearby settlements. A few centimetres might not sound like much, but in a hundred years it could determine the fate of humanity.

BLACK HOLES AND CLIMATE CHANGE

Measuring such small changes is very difficult, which is why scientists are seeking the assistance of an unexpected object: black holes. When I first heard about this, I was astonished. How can black holes in distant galaxies help us understand our own planet? To find out, I travelled to Onsala Space Observatory, about an hour's drive south of Gothenburg on the west coast of Sweden.

WOODPECKERS AND RADIO TELESCOPES

'NO UNAUTHORIZED ENTRY'

I read the sign as my guide drives us through the observatory gates. We pass a couple of grazing deer and a tractor parked by a barn. It looks like an ordinary Swedish agricultural landscape – were it not for the gigantic telescopes peeking out from behind a cluster of trees.

Onsala Space Observatory is operated by Chalmers University of Technology. The observatory's oldest telescope was inaugurated in 1964. At twenty-five metres across, it was one of the biggest telescopes of its time. Swedish astronomers had long dreamed of using a telescope of this size to gain insights into the electro-magnetic interactions between atoms and molecules far out in the coldness of space.

My guide tells me to turn off my phone because the radio signals can interfere with the sensitive measurements. We pass a patch of woodland and I see the Kattegat strait open up on the horizon. A number of sun-drenched telescopes are planted firmly along the low granite cliffs. The pale-blue sky arches over them. This is a place where the land, the sea, the sky – and the birds – meet. The observatory site is a popular haunt for lapwings and both black and green woodpeckers.

FACING INFINITY

'They're a wonder to behold,' says Rüdiger Haas, gazing out towards the telescopes. He's a professor of space geodesy – the art of determining positions and distances on the Earth's surface with the help of objects in space. We meet outside a red wooden cabin that houses a control room. From here it's possible to map our very own planet using supermassive black holes.

'Just as we have lighthouses for seafarers,' Haas explains, 'we have lighthouses in space that we can use as navigation points to see how the Earth is rotating in the universe.'

The 'space lighthouses' in question are quasars. As we saw in Chapter 9, a quasar can form when matter falls into a black hole. This releases a huge quantity of energy, which gets shot out of the black hole and onwards, out of the galaxy. 'Because quasars are so far away from the Earth, we can use them like radio lighthouses in the universe,' Haas says.

As I look out at the strait, I think about how people throughout history have used space to navigate. If we know the positions of the stars and the Sun, we can plot a course on Earth. This technique has enabled seafarers to traverse the seas and oceans. One of the oldest depictions of this is found in Homer's epic poem the *Odyssey*. As Odysseus roamed about, trying to get home to his wife Penelope at the end of the Trojan War, he was captured by the goddess Calypso on the island of Ogygia. After seven years' imprisonment, he was set free and sailed out on a raft. Thanks to the stars, he was able to navigate, since Calypso 'had told him to make his way over the sea, keeping the Bear on his left hand'.[2] With this constellation to the port side, Odysseus sailed out, reaching land after eighteen days. The stars had shown him the way.

Aside from guiding us over land and sea, the stars have also helped people find their way in space. In 1969 Neil Armstrong and Buzz Aldrin were the first people to land on the Moon. Aldrin

BLACK HOLES AND CLIMATE CHANGE

steered a little lunar lander towards the Sea of Tranquillity, the large, dark region of basalt that the astronomers of ancient times mistook for a lunar ocean. Aldrin could see the stars Rigel, Capella and Gamma Cassiopeia through the telescope on the lunar lander, and, by entering the stars' position into a computer, was able to plot a course with a star chart and successfully navigate towards the Moon's surface.[3]

The stars help people navigate and determine their positions on the Earth's surface because they appear to be immobile in the sky. The basic rule of measuring movement is to compare it with something that is not moving. With the help of the stars, it has been possible to study the Earth's surface and learn more about its shape (for instance, that the Earth is flatter at the poles) and rotation (for instance, that the direction of its axis slowly changes over time).

But for Rüdiger Haas and his colleagues, stars have one major limitation: they too are moving through the Milky Way, which means they are not perfectly reliable reference points in the sky. In order to determine precise positions and movements on the Earth's surface, astronomers therefore need to use something that's even further away and more immobile than the stars. 'We need a fixed point of reference,' Haas says, adding, 'The best reference we have is the universe.'

Working with radio telescopes around the world – in Chile, Brazil, Antarctica, New Zealand, South Africa and China, to mention a few – Haas and his colleagues observe the faint radio signals from quasars. These have travelled for many billions of years through space before reaching the telescopes at Onsala. 'We can observe quasars at any time,' Haas says, referring to the fact that radio telescopes can observe signals from space even during daytime. 'When it's raining, when it's snowing, when the Sun shines. They're always there.'

The quasar signals might have been created before the Earth even existed, but today they play an important role in studying our planet. The radio astronomers use atomic clocks to measure the precise moment the quasar's signal reaches each telescope. They enter the measurements into supercomputers that can calculate, with a high level of precision, the location of the quasars in the sky and the location of the telescopes in relation to each other. The technique is the same one used by the Event Horizon Telescope to depict the shadows of black holes: very-long-baseline interferometry (VLBI). In that case, scientists needed to ascertain with very high accuracy the telescopes' positions on the Earth's surface in order to study black holes. At Onsala Space Observatory they need to determine the positions of the quasars with very high accuracy in order to study the Earth. Black holes contribute to our understanding of the Earth, and the Earth's rotation contributes to the virtual telescope that's used to study black holes.

With these quasar observations, radio astronomers are creating a kind of map of the sky called the International Celestial Reference Frame (ICRF). This quasar chart specifies the positions of more than 4,500 quasars.[4] 'The ICRF is our frame of reference for outer space,' Haas says. Just as wayfarers once used star charts to set a course, radio astronomers can establish a telescope's position using these quasar charts. The thing that's special here is the precision: it is possible to determine the position of the telescopes in relation to one another down to the centimetre. It is thus possible to follow, in detail, how fast the Earth is rotating, how the continental plates are moving, and how the land mass upon which the telescopes stand is rising or sinking.

The first ever measurement of continental drift was made with the help of quasars in the early eighties. Over five years, one of the Onsala telescopes, along with the Haystack Observatory and Westford Radio Telescope in Massachusetts, observed around

BLACK HOLES AND CLIMATE CHANGE

fourteen quasars.[5] In 1986 radio astronomers could confirm that the telescope at Onsala was moving away from the telescopes in the US by almost two centimetres a year.[6] The conclusion? That Europe and North America are moving apart at a speed roughly equivalent to the rate at which your nails grow. Today, regular quasar observations make it possible to study how all the continents are moving, tracking them as they move a few centimetres each year. The measurements serve as a complement to several other methods of studying the Earth's surface, primarily measurements via satellite.

The theory of continental drift has been around for a while, but the Onsala telescope observations provided the first opportunity to measure this movement. It was possible thanks to quasars and their power source: supermassive black holes. Surprisingly, research into the Earth's structure and black holes are connected in other ways too.

A SHARED SCIENTIFIC HISTORY

The history of the science of black holes and the Earth are peculiarly interconnected. The rector John Michell was just as curious about the structure of the Earth as he was about the structure of space. As I stated earlier, he was a member of the Board of Longitudes, which was responsible for evaluating methods to determine the longitudinal position of ships at sea.

But Michell also wanted to know what happened beneath the Earth's surface. After a disastrous earthquake in 1755 – which, among other things, destroyed the city of Lisbon and took the lives of up to fifty thousand people – he wanted to know how earthquakes spread through the interior of the planet. He presented the correct hypothesis, that they spread, like waves, from

230   FACING INFINITY

an epicentre. Since then, Michell has been known as the 'father of seismology'.

He also invented an intricate instrument for measuring the Earth's mass. He wrote to his friend Henry Cavendish, 'I have it in contemplation, & hope to try the experiment of weighing the world in the course of the Summer, but wont promise too much for fear of performing too little.'[7] Michell's comment was a bad omen: he managed to build his instrument but died before he was able to 'weigh the world' with it, and his project was completed by Cavendish. It involved taking precise and painstaking measurements of the gravitational effect between two balls and the Earth. Cavendish reported that the density of the Earth was 5.48 times that of water. It was the first time the Earth's density had been estimated with such precision, and the result was surprisingly close to the modern measured value of 5.52.

Another interesting link to the study of the Earth arises in connection with Karl Schwarzschild's formula for a black hole.[8] As discussed, in the autumn and winter of 1915, when he discovered the formula, he was a lieutenant in an artillery unit, and was stationed in the town of Mulhouse, close to the German–French border. Although he devoted his spare time to his astronomical investigations, his days were spent on ballistics calculations for *Langer Max*, the German army's heavy new artillery gun. Over Christmas, many of his superiors went away. 'The loneliness was just starting to feel like too much of a good thing,' he wrote to his wife Else, 'when Alfred Wegener, whose brother is the director of the weather station, turned up.'[9]

Alfred Wegener is today thought of as one of the twentieth century's most important contributors to our understanding of the Earth. But he wasn't a geologist. Wegener had trained in a number of scientific disciplines, initially studying astronomy before writing to his wife that 'astronomy offers no opportunity

BLACK HOLES AND CLIMATE CHANGE

for physical activity'.[10] For this reason he switched to meteorology and headed off on an expedition to Greenland, conducting atmospheric experiments using balloons. When the war broke out, Wegener was drafted and served on the Western Front. He met Schwarzschild in Belgium in May 1915, where they investigated geodesic measurement techniques. While they strolled through the Bois de la Cambre forest, outside Brussels, Wegener talked about his scientific ideas with Schwarzschild, who related to his wife that they were 'fantastical but very stimulating'.[11]

One of these ideas would come to revolutionize our understanding of the Earth. At the time of the First World War, there were many unanswered questions about the history and structure of our planet. No one knew for certain what kind of matter lay beneath the Earth's crust, and whether its surface was fixed or mutable.

Wegener was convinced that the Earth's surface changed over long geological periods. When he looked at an atlas of the world, he was struck by how neatly the east coast of South America fitted the west coast of Africa. Could it not be possible that the two continents were once adjoined? As far back as 1596, the cartographer and geographer Abraham Ortelius suggested that the continents had once been positioned close together before separating. The question was raised several times through the centuries, but there was never any conclusive proof.

'This is an idea I'll have to pursue,' Wegener wrote to his wife in 1910. He observed that there were mountain ranges in South Africa and Argentina, and in Canada, Scotland and Ireland, that had similar structures – which was to be expected if the mountains had been created in the same place. What's more, fossils in western Africa and Brazil, for instance, looked remarkably similar despite the great distances between them. Of particular note were the fossils of the fern *Glossopteris*, which turned up in Antarctica, South America, Australia and India. It seemed natural for this fossil to have appeared in these far-flung places if it had been deposited when the continents were joined together.

Wegener thought that this, and other evidence, indicated that the continents were like a newspaper that had been torn into several pieces. Not only did the edges of the pieces fit together when placed next to each other, their content did too. This convinced him that the continents were moving. He imagined that hundreds of millions of years ago there had been a single, gigantic protocontinent, which he named Pangaea. This was surrounded by a proto-ocean with the poetic name Panthalassa.

In 1915 Wegener published his ideas in the book *The Origin of Continents and Oceans*. At the end of that year he met Karl Schwarzschild in Mulhouse. In a letter to his wife, Schwarzschild described how he and Wegener played billiards and spent more than twelve hours together. All the while, they heard the war rumbling on around Hartmannswillerkopf, the mountain near Mulhouse where French and German soldiers were fighting. Wegener told Schwarzschild about one of his expeditions to Greenland. It's not hard to imagine him discussing his new continental drift theory too. Perhaps Schwarzschild also talked about his new model of the gravitational field inside and around a star. After all, Wegener was originally an astronomer.

At that billiard table a meeting took place between two scientists who hit upon two of the twentieth century's most revolutionary scientific ideas. But neither Schwarzschild nor Wegener lived to see the enormous impact of their discoveries. Schwarzschild died five months after their encounter. Wegener's ideas met with resistance from many geologists. His critics thought that he arbitrarily selected facts that supported his own theory. Sure, there were similarities between the coastlines,

BLACK HOLES AND CLIMATE CHANGE

233

fossils and geological strata, but could these similarities not be explained with the help of geological processes other than the movement of the continents? And where did these great forces come from that were moving land masses about? Wegener never found a satisfactory answer to the latter question. As a result, one geologist asserted that continental drift theory was not a scientific hypothesis, but merely 'a beautiful dream'. Another geologist expressed himself less poetically, dismissing Wegener's theory as 'delirious ravings'.

Wegener died tragically on an expedition to Greenland in 1930. But the criticism of his theory lived on. Interestingly, Einstein supported this criticism. One of the last texts he published before his death in 1955 was a foreword to the book *Earth's Shifting Crust* by Charles Hapgood. Hapgood, a geologist, thought Wegener was wrong. The Earth's surface was changed not by continental drift, but by sudden displacements of the Earth's crust and rotational axis. The shifts would, according to Hapgood, cause catastrophic climate change.[12] Einstein wrote that if Hapgood's theory proved to be correct, it would be 'important to everything that is related to the history of the earth's surface', adding that the theory 'electrified' him.[13] Hapgood's theory turned out to be wrong, but it hasn't been forgotten. In the apocalyptic Hollywood film *2012*, the theory appears as an explanation for the enormous cracks that have opened up on every continent. Ultimately, the theory Einstein had supported became a Hollywood fantasy rather than an accepted scientific explanation.

Einstein doubted both the existence of black holes and the theory of continental drift. Werner Israel, a scientist who, from the mid-sixties onwards, made several significant contributions to our understanding of black holes, believed that there was a psychological reason why these two ideas were met with such suspicion.[14] In both cases, the theories 'each threatened man's instinctive faith in the permanence of matter'. From our short-term human perspectives, it seems the stars will shine forever and the continents will always be in the same arrangement. But everything changes. The continents do move. The stars will one day go out, and some of them will collapse into black holes. The realization that everything changes and that no structures are permanent sweeps away a fundamental belief in the stability of our world. For human beings, both black holes and the shifting positions of the continents mean that we are forced to accept that the world we live in is not static and eternal, but dynamic and changeable.[15]

Interestingly, evidence for and acceptance of these ideas grew simultaneously in the early sixties. At the bottom of the ocean, geologists found key evidence that the continents were moving.[16] Using sonar, which had been developed during the Second World War for locating enemy submarines, the geologist Harry Hess had investigated the bottom of the Atlantic Ocean, finding channels, volcanoes and an extensive mountain range that proved to be the biggest in the world. Just as with the dismissal of continental drift theory, the discovery of the Mid-Atlantic Ridge was met with scepticism: in the fifties, when the geologist and cartographer Marie Tharp argued for its existence, her arguments were initially shrugged off by her boss as 'girl talk'.

Harry Hess realized that new matter was being pushed up from the Earth's interior and was flowing out of the mountain ridge. While this was happening, older material was sinking back into the interior. Hess's discovery, which he published in a revolutionary article that he called 'a geopoetic essay', led to the understanding that the continents rest on large continental plates that are largely covered with sea. As these plates move, the continents follow along in a process known as plate tectonics. Further studies of earthquakes, magnetic traces in geological

BLACK HOLES AND CLIMATE CHANGE

stratifications and structures in the Earth's crust supported the idea of moving continental plates. The geologist John Tuzo Wilson, a contributor to this new theory, confirmed that the world was 'a living, moving thing'.[17]

It was the grip of gravity that ultimately drove the movements of the continental plates, in part by dragging old plate material down into the Earth at subduction zones. At the same time as geologists were observing traces of this enormous force in the depths of the sea, astronomers were observing traces of the huge energy reserves, in the quasar 3C 273 and from Cygnus X-1, that were released as matter whirled around and fell into black holes. Technology that had been developed for military purposes during and after the Second World War – such as sonar, radio measurements and X-ray satellites – made possible observations that led to these theories gaining increased recognition.

But, still, no one had actually measured the movement of land masses. The evidence was indirect. Wegener had hoped that astronomical observations would make it possible to measure the shift in the continents. He himself tried to analyse stellar observations carried out in Greenland and in mainland Europe to find out how the two land masses moved in relation to one another.

It wasn't until 1984 that Wegener's wishes were fulfilled, thanks to the observations of quasars by one of the Onsala telescopes in Sweden and the Haystack Observatory and Westford Radio Telescope on the US east coast. Schwarzschild's and Wegener's discoveries had come full circle. The supermassive black holes behind the quasars had facilitated the direct measurement of the movements of the continents. I can't think of a more striking example of how our knowledge of our planet and our understanding of black holes have developed in parallel.

TIMEKEEPING ON A ROTATING EARTH

'We've learned a lot,' Haas replies when I ask what new knowledge scientists have gained from the quasars' system of cosmic coordinates. 'For example, we've learned a lot about the Earth's rotation.'

As the Earth turns, day becomes night and night becomes day. In fact, this regularity makes the Earth a gigantic clock that provides us with our most important unit of time: the day. But the time it takes for the Earth to complete a full rotation can vary. 'The Earth is not a solid body,' Haas explains, 'it's made up of different parts. There's the core, the asthenosphere, the lithosphere, the crust, the oceans, the atmosphere and so on. These parts are constantly interacting.'

As rotational energy flows between these parts, the length of the day changes. Every time the El Niño weather system sweeps down into the southern hemisphere, for example, the atmosphere moves faster.[18] The change in energy causes the Earth to rotate more slowly, making the day longer. Although the variation is small – less than a millisecond – Haas and his colleagues can measure it with the aid of VLBI and quasars. Since measurements began more than six decades ago, the cumulative effect of El Niño on the Earth's rotation has led to the day getting longer by less than a tenth of a second – less time than it takes to blink!

But in the longer term the planet's rotation will decrease dramatically. 'We've learned that the Earth is spinning slower and slower over time,' Haas says. The Moon's gravity, and to a certain extent the Sun's, causes the oceans' tides to sweep in and out. The movement of these enormous masses of water, the friction on the seabed, and the gravitational interaction between the Moon and the Earth lead to a reduction in the speed of the

BLACK HOLES AND CLIMATE CHANGE

Earth's rotation by approximately 1.8 milliseconds per century. Over several million years this change becomes significant. When Tyrannosaurus Rex walked the Earth, the day was half an hour shorter than it is today.[19] In 200 million years, a day will be approximately twenty-five hours long.[20] The gravitational interplay between the Moon, the Earth and its oceans is also causing the Moon to move four centimetres away from Earth each year. In the far future, about 600 million years from now, any human who is still around will therefore be unable to enjoy total solar eclipses, because the Moon will no longer be the same size in the sky as the Sun.

Climate change is also affecting the Earth's rotation. As the ice at the poles melts, bodies of water are moving in such a way that the Earth's rotation is slowing. An international team of researchers has calculated that melting pack ice may lead to a decrease in the rotational speed of 2.62 milliseconds per year.[21] This is more than the equivalent effect from the Moon.

When I hear about these small changes in the Earth's rotation, I think they sound pretty immaterial. But Haas tells me that it's important for the functioning of our society that we measure the changes. Bank transactions, scientific experiments, satellites and other high-tech applications require time to be measured in tenths of a millisecond, so our measurement systems must be capable of such accuracy. Today we use two main internationally established systems of time. One is based on the Earth's rotation, the other on an international network of atomic clocks.[22] These two time standards have to be synchronized regularly as a result of the small shifts in the Earth's rate of rotation. Because quasars help us keep track of this rate, black holes actually contribute to our timekeeping. Given that they are distant objects where space and time apparently cease to exist, it is rather ironic that we use them to study space and time here on Earth.

In the future, Haas and his colleagues want to use quasars to measure the varying rates at which the clocks at different telescopes tick.[23] This not only involves measuring shifts in the Earth's surface, but changes in the passage of time in different places on Earth. Quasars will soon enable us to map not only the Earth's physical space but its spacetime too.

Quasars also contribute to navigation on Earth and in space. When organizations such as NASA and ESA send spacecraft to other planets, moons and asteroids, they have to know the precise position and speed of those vessels.[24] This can be established by comparing the position of the spacecraft with the locations of quasars in the sky. One space telescope has even helped scientists pinpoint the movement of the solar system with the help of quasars.[25] Using the space observatory Gaia, scientists have mapped more than 1.6 million quasars. By analysing their light, scientists were able to determine precisely how our solar system is accelerating towards the centre of the Milky Way. It is this acceleration that creates our galactic orbit, just as the Earth's orbit is a result of its acceleration towards the Sun. Quasars enable us to measure the Earth, follow trajectories within the solar system and chart the solar system's course through the universe.

Global navigation systems such as GPS are calibrated using quasars. The precision of global navigation satellites has constantly improved, and today they can be used to determine positions on the Earth's surface to within a metre. This requires precise knowledge of the positions of the Earth and the satellites. 'The people managing global navigation systems need information about the rotational angle of the Earth in order to calculate satellite trajectories as accurately as possible,' Haas tells me. This information comes from quasar measurements.

I watch a lapwing darting across the meadows around the telescopes. It has no idea they are gathering radio signals from

BLACK HOLES AND CLIMATE CHANGE

black holes, enabling us to understand our planet and our place in the world. Until I heard about the research being conducted at Onsala Space Observatory, I was as unaware as the lapwing that this was possible.

I ask Haas whether it's accurate to say that black holes help us navigate on Earth.

'Yes, it is,' he replies.

'Isn't that crazy?' I hear myself exclaim.

'Depends what you mean by crazy,' Haas replies, smiling. 'You have to get used to the thought. It's not crazy but it is incredibly fascinating.'

DANGER ON THE FALSTERBO PENINSULA

But what about the threat from the sea?

As well as determining the Earth's rotation, Haas and his colleagues can see how the relative positions of the telescopes on the Earth's surface change over time. 'We can see the effects of land rises and tectonic movements,' Haas says. 'And then we can predict when a station is at risk of flooding.'

Roughly three hundred kilometres south of Onsala, on the south-western tip of Sweden, lies Skanör-Falsterbo. It's a prosperous county on the Falsterbo Peninsula, flanked to the north by the Øresund strait and to the south by the Baltic Sea. In 2017 a storm swept in and the sea level rose by 1.5 metres. Cellars, golf courses and the coastal meadows of the Flommen nature reserve were submerged. When another storm blew in, during the autumn of 2023, the local authorities were prepared, with ramparts and pumps protecting the community from the high waters.

But in the long term, the whole of the peninsula is at risk of being drowned by the waters that surround it. Whether or not this happens will depend on how Sweden's land mass changes in relation to rising sea levels. Approximately twelve thousand years ago, the northern hemisphere was covered with glaciers. As they melted away, the parts of the Earth's crust that had previously been pressed down beneath their weight began to rise up. It's like pressing down on a mattress and then taking your hand away: after a while the mattress returns to its original shape. Many land masses are still rising in this way. The change in sea level along a given stretch of coast is therefore dependent on how much the water level and the land rise in relation to each other.

The Scandinavian land mass has risen by up to three hundred metres since the last ice age. But this rise had happened faster in northern Sweden than in the south. In the south, the sea level is rising faster, meaning that in 250 years large parts of Skåne, the southern-most region, could be under water. A similar north–south division occurs with the British Isles, where the land mass of Scotland and northern England has been rising steadily since the last ice age, whereas southern England and Ireland are sinking on the order of ten centimetres per century. In North America, this post-glacial rebound is mostly causing land rise in parts of Canada and the northern US, with the US east coast and gulf coast instead being most at risk from rising sea levels.

I ask Haas if he would recommend that the Skåne residents in southern Sweden up sticks now and move to more northerly regions. 'They can if they want,' he says with a laugh, 'but it's a little colder there.'

Behind the light-hearted tone lie serious concerns. Since the beginning of the twentieth century, sea levels have risen by an average of twenty centimetres.[26] By about 2050 it is possible that they will rise another thirty centimetres. However, the IPCC warns that they could rise by as much as a metre by the end of the century, and melting ice caps could contribute an

BLACK HOLES AND CLIMATE CHANGE   **241**

additional metre. There is huge uncertainty. The most significant factor that will influence sea level rise is the amount of carbon dioxide released into the atmosphere. Much of the damage has already been done, and even if carbon dioxide emissions were to reduce drastically, the current warming of the world's oceans will continue to have consequences far into the future. Aside from vast areas of land being submerged, storms – like those that have struck the Falsterbo Peninsula – will increase in frequency and intensity, causing recurrent damage. To prepare for the catastrophes, politicians, scientists, urban planners and citizens themselves need to know how much sea levels will rise in different locations. Quasar measurements will therefore be required, since they enable scientists to find out how land masses around the world are rising or falling.

The terrifying climate changes we are witnessing are a stark contrast to the tranquillity of the landscape around the Onsala telescopes. As I stand by one of these enormous machines and look out across the rocks, I think about how humans are capable of measuring a centimetre's shift in a continental plate or a few nanoseconds' reduction per day in the time it takes the Earth to spin on its axis. I'm struck by how, despite this detailed understanding of our planet, we're unable to take care of it. What's the point of all this knowledge and technology if we can't look after our world, our only home in the universe?

A BLUE DOT

My visit to Onsala is over. I bid farewell to Rüdiger Haas and get a lift back to Gothenburg with my guide. As I pass the gates on the way out of the observatory I turn my phone back on, and it once again starts to pick up signals from GPS satellites and radio masts. I open a navigation app and look at the blue dot telling me I'm outside Onsala. I watch my position updating as the car travels towards the city.

After my trip to Onsala, I no longer have quite the same image of this little blue dot. It's not only a symbol that indicates where I am, it's also an example of how humans are able to utilize black holes in distant parts of the universe. Quasar observations play a small yet practical role in our everyday technology. The name Pōwehi – 'the embellished dark source of unending creation' – really did suit a black hole. In the case of quasars, they create a never-ending radio signal that we can harness for technological applications on Earth.

Quasars help us understand how the Earth's surface will change in the future. But can black holes also help us understand our past? Thanks to new observations, some scientists think that could be the case. They've started to realize that black holes may have played a part in both the emergence and the evolution of life on Earth. Perhaps our very existence was made possible by black holes. After my trip to Onsala I board a train bound for the Netherlands to find out how.

Chapter 12

· · · · · · · · · · · · · · ·

BLACK HOLES
ARE OUR FATHERS

'The cosmos is within us. We are made of star-stuff.'
This quote from *Cosmos*, an eighties TV documentary by the prominent astrophysicist Carl Sagan, has become very famous. The realization that almost all the atoms in our bodies were created from stars that burn, explode and collide made it possible to answer one of humanity's oldest questions: where does everything come from?

'The nitrogen in our DNA,' Sagan went on, 'the calcium in our teeth, the iron in our blood, the carbon in our apple pies were made in the interiors of collapsing stars.' But in recent decades scientists have become aware that our origins are to be traced back not only to the stars but also to black holes, which seem to have played a surprising and important part in the inception of life on Earth – and perhaps throughout the universe.

At SRON Netherlands Institute for Space Research in Leiden, I meet the astrophysicist Aurora Simionescu. I contacted her after watching a documentary in which she said, 'Supernovas are our mothers, but on the other hand, black holes are our fathers.'

'First I was in a Japanese documentary about black holes, and they wanted me to say it,' Simionescu tells me. 'Then a German team saw it and wanted me to say the same thing in their documentary, so I've said it twice!'[1]

Simionescu is in her forties, full of energy, and as she talks she gesticulates enthusiastically. She has just moved into SRON's new premises, where they build space instruments for NASA, ESA and their Japanese equivalent JAXA. The instruments find their way onto satellites that are used to study everything from the Earth's climate to the origins of the universe.

As we walk through the building we pass lots of small models of satellites. Simionescu tells me that, despite the beautiful new facilities, she does her best thinking in her free time, while diving or dancing. 'Staring at a computer screen all day doesn't give you new and fresh ideas,' she says. 'Even if I have a problem to solve at six o'clock on a Friday night and I have a dance class booked, I go and dance Brazilian zouk. My brain keeps thinking about the problem, but if I change the environment, I analyse it from a different angle.'

One of the problems keeping her busy is the question of how matter is created and distributed to different places in the universe. As we saw in Chapter 6, at the end of its life a star can explode into a supernova. When this happens, new elements are both created and scattered into space. 'One variant happens when a star with a large mass collapses, and another when a white dwarf collapses,' Simionescu explains. 'They create very different substances. When a white dwarf explodes, it is completely blown to pieces. This creates very heavy elements like iron and nickel. But a very high-mass star has a core of iron, and when the star explodes, the core is compressed and becomes a black hole or neutron star. It's only the outer layers of the star that are scattered in space, and those layers contain lighter elements, like oxygen and magnesium.'

In this celestial alchemy, the newly created elements are flung out at incredible speeds. They travel through the space between the stars, sometimes ending up in new stars or planets. That's why supernovas are 'our mothers' – they have created much of the matter we are made of.

When I enter Simionescu's office, I see a whiteboard covered in scrawled equations. Through the window I can see a few trees and university buildings in the dreary autumn weather. When I look at the trees I think about how some of their atoms were once formed in an exploding star. Perhaps this is one of the deepest insights astronomers have had: that everything is part of a great cosmic cycle. The German-US physicist Hans Bethe, who was awarded the 1967 Nobel Prize in Physics for his work on the energy production in stars, painted a beautiful picture of this cycle: 'Stars have a life cycle much like animals,' he said in his Nobel Lecture at Stockholm City Hall. 'They get born, they grow, they go through a definite internal development, and finally they die, to give back the material of which they are made so that new stars may live.'[2]

Since some of the stars that exploded into supernovas became black holes, some of the elements on Earth must have been created in the same process. For instance, some of the oxygen we breathe could have been ejected by a collapsing supernova just before the formation of a black hole. 'Carl Sagan said we are stardust,' Aurora Simionescu says, 'but we also breathe stardust. *Stardust that was partly helped by a black hole.*' These black holes live on in the Milky Way as a dark reminder of our cosmic origins.

FLOUR, SUGAR AND EGGS IN EVERY CORNER OF THE UNIVERSE

In 2012 Simionescu made a discovery that demonstrated how even supermassive black holes can affect life on Earth. At that time she was the Einstein Fellow at the prestigious Kavli Institute for Particle Astrophysics and Cosmology in California. Her work involved analysing data from the X-ray telescope Suzaku, named after a majestic red bird in Asian mythologies. The telescope was the result of a collaboration between NASA and JAXA.

Simionescu's boss Steven Allen suggested she use Suzaku's observations to identify which elements might be found in the regions between galaxies. There are several smaller galaxies orbiting the Milky Way, and this galactic family, known as the Local Group, also encompasses the more distant Andromeda galaxy. This group is not isolated, however. Other galactic groups form gravitational kinships, coming together in larger communities known as clusters that may consist of thousands of individual galaxies. Our Local Group is close to the Virgo cluster, fifty-five million light years away. At the centre of this cluster is the giant galaxy M87, which houses the supermassive black hole Pōwehi. 'The galaxies in the universe are not randomly scattered,' says Simionescu. 'There are places where galaxies are packed together, and there are long filaments and voids between them. In the densest parts of this cosmic web, in the galaxy clusters, the stars contain about 10 to 15 per cent of all normal matter. The rest is scattered in the void between galaxies.'

The region between the galaxies is known as the *intergalactic medium*. On average it contains just one atom per cubic metre of space. But since the distances between the galaxies are so huge, the sum of all the atoms in these cubic metres represents an enormous quantity of matter. Therefore the combined mass in the intergalactic medium can be greater than that in the galaxies themselves. As Simionescu points out, 'Most of the atoms in the universe will never end up in a star or a planet.'

I ask her how she pictures the immense space between the galaxies. 'I try not to imagine it,' she replies, 'because my brain explodes. I'm often asked if I feel insignificant compared to these enormous distances in space. But then I answer that I feel huge, because even though my body is small my mind can explore these distances.'

The Suzaku telescope had picked up X-ray signals from various parts of the Virgo and Perseus clusters. The latter is more than four times as far away as the former, and contains many thousands of galaxies.[3] In 2022 the supermassive black hole in the middle of the Perseus cluster's central galaxy became famous as a result of its terrifying noise. Matter is expelled from the area surrounding the black hole at such regular intervals it creates vibrations in the hot gas of the intergalactic medium. Just as the sounds we humans can hear are vibrations in the air, these cosmic vibrations are actually a kind of space noise. It would be impossible to hear these sound waves with our ears, because they have a frequency fifty-seven octaves lower than middle C on a piano. The NASA scientists translated the X-ray signals they had picked up from these plasma waves into a sound the human ear can hear.[4] This sonification sounded like a demonic choir singing in the intergalactic medium, which led to it going viral on the internet.

What Aurora Simionescu wanted to know, however, was which elements could be found in the intergalactic medium. The X-ray signals from galaxy clusters contain a kind of fingerprint of the elements in the hot but thin plasma between the galaxies. When Simionescu analysed the data from the Suzaku telescope, it provided a new view of the role black holes can play in the cosmic cycle of the elements. 'I was completely floored by what I saw,' she says. In the intergalactic medium, she saw traces of different elements such as iron, magnesium and silicon.[5] What astonished her was the relative abundance of these elements.

'Iron, magnesium, silicon sulphide and other elements occur in exactly the same ratio as in our own solar system,' Simionescu explains. 'There's no reason for the composition of the solar system to be the same as a galaxy cluster fifty-five million light years away. It felt like a big cosmic conspiracy.' Some unknown process was ensuring that the elements created as stars burned, died and occasionally collided were being mixed and scattered in even proportions, not only within galaxies but between them too. Simionescu had one main suspect: the supermassive black holes that are found at the centres of most galaxies. 'Among the general public,' she says, 'a common belief exists that black holes suck in all matter around them. But when I give popular science talks, I usually say that black holes don't suck. What we have discovered is that the most important action of black holes is that they blow things away. If they didn't send matter out of galaxies, the universe would look very different.'

When a supernova explodes, the newly created elements are flung out into the galaxy. But the galactic supernovas rarely produce enough energy to hurl atoms with the immense force required for matter to leave the galaxy. This is only possible thanks to the strength of the jets supermassive black holes produce. As we saw in Chapter 8, these jets are produced by a complex interplay of the black hole's gravitational force, rotation and magnetic field. They send out light and matter to distances on a completely different scale to the galaxies they are found in. In 2024 a group of astronomers announced that they had observed a jet that was twenty-three million light years long, possibly one of the largest structures ever created from a galaxy.[6] The jet was initially detected when both machine learning algorithms and citizen scientists combed through the vast quantities of data collected by the global radio network LOFAR. Because of its enormous size, the jet was dubbed Porphyrion, after a giant in Greek mythology. The leader of the research team, Martijn Oei from Leiden University, stated that its huge size

BLACK HOLES ARE OUR FATHERS

made it clear that jets expelled by supermassive black holes can affect 'every place in the universe... at some point in cosmic time'. It is probable that jets like Porphyrion were once more common, and in 2025 a group of astronomers reported that they had indeed observed a record-breaking jet, twice as long as the Milky Way, in the early universe.[7] 'When the universe was younger,' Simionescu explains, 'it was easier for matter to get kicked out from galaxies. Galaxies weren't fully formed back then and their gravitational pull wasn't as strong as today.'

As elements find their way into the intergalactic medium they mingle at timescales beyond those of human existence. Some of the elements later fall back into the galaxies. 'The beautiful thing,' Simionescu says, 'is that because the mixing happens in the void between galaxies, and galaxies grow from the material around them, it's entirely possible that some of the atoms in your body come from another galaxy.'

After I hear her striking assertion, I try to imagine the cosmic journey of one such atom. Perhaps a supernova explosion in a distant galaxy tossed out an iron atom several billions of years ago. The atom was propelled into space and found its way into another star that grew and eventually threw off its outermost layers. The atom floated about in the galaxy, got close to the supermassive black hole at its centre and was spewed out by the enormous force of the black hole's jet. Then the iron atom began its several-billion-year-long journey through the intergalactic medium, where it was eventually caught by the gravitational hold of the Milky Way. It continued into the interstellar space of our galaxy and after travelling between the stars for millions of years it and other atoms ended up on the edge of our solar system, from where it was pulled closer to the Sun by a comet. The comet hit Earth, and after several transmutations in minerals, plants and animals, the iron atom finally found itself in my blood. Its

250 FACING INFINITY

intergalactic journey has taken it all the way from a supernova in another galaxy to the office in Leiden where Aurora Simionescu and I are discussing the cosmic origins of atoms just like it.

'There is a poetry in how everything is connected,' she reflects. 'You can't say "I'm just going to study the Milky Way", because there are processes in the Milky Way that are part of a larger cosmic environment. Everything is connected. In a way, I think that's more beautiful than scientific.'

We say that everything on Earth is interconnected, but the same is true on a cosmic scale. 'We now understand that the evolution of the galaxy depends on what the black holes do,' says Simionescu. 'A galaxy is not an isolated system. Supernovas explode, supermassive black holes send matter out of the galaxy, and gas flows back from space between galaxies. I believe it is this cosmic circulation that mixes all the matter in the universe.'

This cycle in which black holes play an important part could be the reason for the ratios of elements in the intergalactic medium being the same as in our solar system. 'Suppose you want to bake a cake,' says Simionescu. 'Then you need the right proportions of eggs, flour, sugar, butter and so on. But that by itself doesn't mean the cake will get baked. Similarly, the right proportion of carbon, magnesium, silicon and so on in different places in the universe doesn't guarantee that life will automatically emerge. But metaphorically speaking, black holes ensure that every corner of the universe has the right amount of flour, sugar and eggs. This is one of the conditions for life.'

She emphasizes that this is a theory that needs to be tested more carefully with new observations and improved models.[8] She also says that much of it is speculative, since right now there is only one place in the universe where we know life exists – Earth. 'Life might look very different in other places in the universe,'

BLACK HOLES ARE OUR FATHERS 251

she says. 'There are likely to be additional factors for the origin of life that we don't understand. But what I argue is that black holes ensure that the basic condition is there: the presence of the right ingredients.'

Black holes could thus play a role in the creation, and the spreading, of some of life's building blocks. In this sense, black holes are our fathers. But they can also play a destructive role, something that became clear one autumn day in 2022.

GAMMA-RAY BURSTS AND MASS EXTINCTION

On 9 October 2022, the Earth's atmosphere was subjected to an intense bombardment of X-rays and gamma rays.[9] In a few seconds the photons deposited as much energy as ten million lightbulbs. Measuring stations around the globe captured the effect of the atmospheric disturbances on radio communications on Earth.[10] Even lightning detectors in India were triggered by the energy discharge.[11]

Using the Neil Gehrels Swift Observatory in orbit over Earth, astronomers determined that the X-rays came from the constellation Sagitta, and other telescopes helped identify the source of the gamma radiation. The gamma-ray burst was caused by an explosion in a distant galaxy 2.4 billion years ago. Its formal name was GRB 221009A but astronomers called it BOAT, 'Brightest of All Time', because it was the most powerful gamma-ray burst they'd ever observed.

These kinds of intense bursts of gamma radiation were first observed in 1967. The US had launched several satellites to monitor whether the Soviet Union was sticking to a new agreement on nuclear weapons testing. The satellites registered powerful bursts of gamma rays, but they came neither from nuclear weapons nor from any other process on Earth. Two astrophysicists at Los Alamos in New Mexico demonstrated that the bursts came not from our solar system, but from objects far off in space.

The origin of gamma-ray bursts is not entirely known. Astronomers believe that they can be created when matter falls at high speed into a black hole, when neutron stars collide, or when a massive star collapses to form a neutron star or a black hole. This last phenomenon usually leads to a supernova explosion, but in rare cases, gamma radiation can be focused at the poles of the collapsed star. When this focused radiation is aligned in our direction, we see it as a gamma-ray burst, in what could be a violent echo from a newly created black hole.

The energy released in a gamma-ray burst is huge. In a few seconds it generates more energy than the hundreds of billions of stars in the Milky Way generate in a whole year. In the case of BOAT, the explosion happened in a galaxy very far from Earth. In spite of that, it showed up as a hiss in our radio communications. In 2024 observations with the James Webb Space Telescope showed that the gamma-ray burst was probably caused by an incredibly massive star exploding into a supernova.[12] Since a black hole was likely the end product of such an explosion, it means that the birth of a black hole in another galaxy can impact radio communications on Earth!

What would happen if a gamma-ray burst originated in the Milky Way? Along with scientists from NASA, the astrophysicist Adrian Melott from the University of Kansas has advanced the idea that it was a gamma-ray burst that caused a significant proportion of life on Earth to go extinct 440 million years ago.[13] At this time, many of the Earth's marine invertebrates died out in the so-called Ordovician–Silurian mass extinction. Biologists have long speculated about what could have caused this event. Perhaps it was increased volcanic activity or a change in the

BLACK HOLES ARE OUR FATHERS

metal content of the Earth's oceans. But Melott and his colleagues suggested instead that it was due to an explosion in our galactic neighbourhood. In just ten seconds the radiation from a gamma-ray burst would have decimated up to half the Earth's ozone layer. This would mean the Sun's ultraviolet radiation would penetrate all the way to the surface of the Earth, at the same time as molecular reactions in the atmosphere might lead to the climate becoming colder and the glaciers growing in size. The combination of the colder climate and the ultraviolet rays would have led to many species dying out.

Melott and his colleagues estimated that if it had been a gamma-ray burst that impacted the climate, the explosion would have to have happened around 6,000 light years away. Even if there is no unambiguous proof that a gamma-ray burst was behind the Ordovician–Silurian mass extinction, Melott's argument demonstrates that the evolution of life on Earth could be sensitive to exploding, black hole-forming stars. The researchers therefore call gamma-ray bursts angled towards the Earth a 'serious threat to the biosphere'. Thankfully, it's extremely rare for gamma-ray bursts to happen so close; approximately two such events happen every billion years.

In 2023 astronomers got new proof that black holes can give rise to some of the most energetic phenomena in the cosmos. They detected a supermassive black hole located more than seven billion light years away. It had torn apart an ill-fated star that had come too close. The matter left behind accelerated to great speeds and emitted so much radiation that the system shone more brightly than thousands of supernovas. Astronomers were so shocked by the quantity of energy they named the object 'Scary Barbie' – a play on the letters in the object's formal name ZTF20abrbeie.[14]

THE GALACTIC HABITABLE ZONE

Black holes may have another role to play in the evolution of life. Astronomers have long wondered whether it is possible for life to evolve anywhere in a galaxy. Space is a ruthless, dangerous place. There are many potential cosmic threats to the life on a planet. The best-known is the asteroid that hit the Earth and caused the extinction of almost all the dinosaurs. But in space there are other, much more dangerous phenomena. 'I would be very worried if a star near the Earth exploded in a supernova,' Aurora Simionescu says.

The red star Betelgeuse, which can be seen in the constellation Orion, is the closest star to Earth that is at risk of doing so. It's impossible to say when it will happen: it might be tomorrow or in many millions of years. If Betelgeuse were to go supernova, it would be visible as a brightly shining point in the sky, so bright it could be seen during the day and would cast shadows at night. But luckily Betelgeuse is so far away it wouldn't pose any risk to our planet.

Exploding supernovas could, however, constitute a threat to life on other planets.[15] Astronomers have therefore asked the question of whether there might be certain distinct regions in a galaxy where life can occur. In these habitable zones, life would be able to develop without the risk of being snuffed out by supernovas and gamma-ray bursts. The idea emerged because there is a habitable zone in our own solar system. The term is defined as the area in which planets around a star can have liquid water on their surface. Earth is in the centre of this zone in our own solar system: there is liquid water and an environment that supports life. Mars lies just beyond the outer boundary and Venus beyond the inner boundary.

It is, however, hard to define exactly where a galactic habitable zone would begin and end. One possible condition is that

BLACK HOLES ARE OUR FATHERS   255

the right elements would have to exist for planets and moons to form, another may be that there cannot be too many stars that explode in gamma-ray bursts and supernovas.

Even if simple life forms do occur, it remains unclear which conditions would lead to them evolving to more advanced levels and even intelligent life. The only data point we have is our own planet. 'Life appeared very early in Earth's history,' says Simionescu. 'The oldest rocks contain the oldest signs of life. But it takes a long time to go from that stage to intelligent life. You need stability. Too many interactions between stars can have a negative impact on their planets.' In areas with many stars – such as the centre of the Milky Way – planets might also be pulled off their orbits as two stars pass close to each other.

Supermassive black holes could therefore have another influence on the development of life, aside from spreading matter within and between galaxies. Because of their jets and other astrophysical processes, black holes can regulate how many stars form in the galaxies they occupy. The energy released from supermassive black holes can heat up and even blow away gas in the galaxy, which prevents the formation of new stars. But it's also possible that the displacement and accumulation of the galactic gas can provide new material for stars, which shows how complex the black hole influence on the evolution of a galaxy can be. If the activity of the black hole leads to a lower degree of star formation, it will mean fewer supernovas and fewer ill-fated encounters between stars, potentially leading to a more stable environment.[16] It is therefore possible for black holes to have an influence over the galactic environment in which advanced, intelligent life can develop.

But it is probably not possible for advanced life to develop too close to the galactic centre. Supermassive black hole jets may also play a destructive role by disrupting the atmospheres of the planets orbiting the stars at the galactic centre.[17] It would be hard, or even impossible, for life to evolve on such planets. But as with the case of star formation, supermassive black holes can have both a positive and negative effect. A study published in 2025 showed that the radiation from an active galactic centre could, under certain conditions, help to create an extra thick ozone layer on a planet that is at a suitable distance to the centre.[18]

Simionescu emphasizes that we do not know exactly what effect the jets from supermassive black holes might have on the possibility of life. 'If you have a lot of radiation from the centre of the galaxy, it can trigger mutations in primitive life forms,' she reflects. 'The radiation might damage the life forms, or it might cause them to evolve. Both options are possible.'

The question is what role Sagittarius A* has played for our galaxy. Astronomers have not observed a jet issuing from it, which means that if there is one, it's very weak. But the researchers from the Event Horizon Telescope team were able to demonstrate that the rotational axis of the black hole was almost pointing towards Earth. The alignment was pure coincidence, but it raised the question of whether a jet from Sagittarius A* could shoot straight into the galaxy. 'It's a quiet black hole,' says Simionescu. 'But it has been active before. Maybe there is a connection between the activity of the Milky Way's supermassive black hole and the evolution of life in the galaxy.'

There's a great deal of uncertainty around this subject. Perhaps our position in the galaxy – far from the supermassive black hole and star-dense regions – has been favourable for life. Since we can't travel around and find out whether life exists on other planets or moons in the Milky Way, it's difficult to exactly delineate the galactic habitable zone and how it might change over time. Perhaps the James Webb Space Telescope will unambiguously spot traces of special molecules in the atmosphere of the planets around other stars, molecules that can be created only by biological processes (a prime target being the exoplanet K2-18b, for which preliminary but contested results have been reported). A decisive finding like that would improve astronomers' estimates as to the prevalence of life in the Milky Way. What is already clear, however, is that astronomers cannot exclude supermassive black holes from their calculations concerning the origins of life in the universe.

LIFE ON BLANETS

Life exists on Earth. There may also be, or could have previously existed, simple life forms on our neighbouring planet Mars. Some of the moons in our solar system could also conceivably accommodate life. Two of the main candidates are Europa, which orbits Jupiter, and Enceladus, which orbits Saturn. Beneath their ice-covered surfaces, these moons have liquid oceans in which life could feasibly evolve.

But might there even be life around black holes? Moons orbit planets and planets orbit stars. Some scientists have suggested that there might also be planets orbiting black holes, and that such planets could even harbour life.

A planet orbiting a black hole is called a *blanet*, a term introduced in 2019 by the Japanese scientist Keiichi Wada and his colleagues.[19] Through astrophysical calculations they showed how blanets could be formed from the gas clouds that surround supermassive black holes, and might be many thousands of times more massive than the Earth. Their number is staggering. According to Wada's calculations, more than a hundred thousand such blanets might be formed from the matter around a black hole, and the crowd of blanets might take anywhere from a hundred thousand to a million years to complete an orbit, since they might be many light years away from it.

In 1992 the first planets beyond our solar system were found. Interestingly, they were orbiting a neutron star. Since a neutron star is the next-most compact object in space after a black hole, it seems likely that planets – or blanets, to use the terms interchangeably – could also orbit black holes formed from imploding stars.

Blanets are rich fodder for science fiction. But could life occur in such extreme locations? In order to form and evolve, life, at least as we know it on this planet, needs not only the right building blocks, but also an energy source. On our planet, this means the Sun and the internal heat energy of the Earth. For a blanet orbiting a stellar-mass black hole, the energy source could be the glowing accretion disc surrounding the black hole.

The Czech physicist Tomáš Opatrný and his colleagues have determined what the conditions for life on a blanet orbiting a supermassive black hole would be.[20] The sky on a blanet would be pitch black. Instead of a shining star, it would be dominated by a dark black hole. One possible energy source that could contribute to the development of life on a blanet would be the cosmic background radiation. This radiation fills the whole of space. It has existed since the formation of the universe and will exist into its distant future. The cosmic background radiation was very hot in the universe's infancy, but is today extremely cold, just a few degrees above absolute zero.

Because of the curvature of spacetime around a black hole, the cosmic background radiation would heat up. As photons travel through a black hole's gravitational field, their energy increases. An observer close to the event horizon would register high-temperature background radiation, the light from which would be concentrated into a small area of space by the black hole's spacetime curvature. On a blanet close to a black hole, the sky would therefore be largely filled by the black hole's frigid darkness, but a small part of it would be constituted by the focused, heated background radiation. On the Earth, the temperature difference between the Sun and the cold void of space is a prerequisite for life. On a blanet it would be the difference between the cold black hole and the hot background radiation that would set the conditions for life's existence. The intense radiation from all the gas and dust orbiting a supermassive black hole would also provide some of the energy required, as well as the radioactive processes and other heat sources that might exist within the blanet.[21] 'But it would be a sad life,' Simionescu reflects, 'since there's no starlight. At the same time, there might be life forms in space that are so strange that we can't imagine them.'

When scientists study extreme environments on Earth, they often discover that they are teeming with life. It makes no difference whether it is incredibly dry, cold, hot, acidic, alkaline, saline or radioactive. Microbes still flourish. Life appears to be sustained in so many different environments that it may be these extreme conditions that are normal and our own circumstances that are the exception.

Life forms on a blanet would also be special, but in a way that is not possible on Earth. They would live in an extreme time environment. Because of the black hole's gravity, time on a blanet would pass more slowly relative to an observer far away from the system. The size of the time delay would depend on the mass of the black hole and the blanet's distance from it. The effect might be very large if the blanet were close to the black hole. A year on its surface might last thousands of years – or even longer! – further away. It's as though the blanet's life forms would exist in a pocket of time, distinct from the rest of the universe. When several million years had passed on the blanet's surface and its peculiar life forms had evolved, several billions of years could have passed on Earth. Equally, our own Sun might have gone out, just as life on one of these planets was beginning to develop. And if these life forms were ever to reach a more complex stage, all the stars in the universe might have already stopped shining. The life forms on the blanet would be able to survive longer than the life that exists around stars, because the cosmic background radiation will continue to exist throughout the universe (even though over time it will grow colder and colder). The possibility of such life forms existing is speculation that borders on science fiction. But I can't help wondering if the last life forms in the universe will exist on a planet that orbits not a star, but a black hole.

FROM CHILDHOOD TO ADULTHOOD

My conversation with Aurora Simionescu is over. We leave her office and walk along the corridors of SRON's newly built home. The walls are hung with posters showing the instruments scientists at SRON are involved in developing and building. One of them is LISA, the advanced space-based observatory that will listen for gravitational waves from colliding supermassive black holes.

Simionescu tells me proudly that a new X-ray telescope known as the X-ray Imaging and Spectroscopy Mission has been launched as planned. She can now use it to study in detail the flow of matter around the gigantic galaxy M87. This will give her a better understanding of the interplay between the supermassive black hole Pōwehi and its cosmic surroundings.

It is common to associate black holes with death, but perhaps we should, instead, be associating them with life. Just as in Hawaiian mythology, where darkness is viewed as a life-giving

BLACK HOLES ARE OUR FATHERS 261

force, we should appreciate black holes for the part they may play in making life possible.

'We now realize that supermassive black holes play an important role in the universe,' says Simionescu. 'Galaxies and black holes grew up together, from childhood to adulthood. Galaxies, galaxy clusters and the cosmic web would all look different had it not been for the action of supermassive black holes. But even though we know that supermassive black holes affect several phenomena in the cosmos, we do not yet know exactly how.'

Research into the relationship between black holes and the origin of life is in its infancy. It is necessarily speculative. After I take leave of Aurora Simionescu I think about how the speculations of today could be the scientific results of tomorrow – and that supernovas are our mothers, and black holes are our fathers.

Chapter 13

· · · · · · · · · · · · · ·

HAWKING'S
LAST JOURNEY

In the spring of 1974 farmers in China discovered an army of terracotta warriors in a field, Stephen King published his debut novel *Carrie*, and at the Brighton Dome on the south coast of England, ABBA won the Eurovision Song Contest with 'Waterloo'. The song's title referred to Napoleon's defeat near the Belgian city of the same name.

That same spring at a conference in Oxford, Stephen Hawking declared that he had made an important discovery related to black holes. Sitting in his wheelchair, Hawking announced, in a weak, indistinct voice, that he had calculated that black holes were not completely black. They could glow, and when they did so, they would lose mass and shrink. Hawking had realized that black holes too will meet their Waterloo, disappearing completely in the end.

Several of the participants at the conference reacted to Hawking's new results with consternation. Some of them did not believe his calculations, and one professor even dismissed them as 'absolute rubbish'.[1] It had taken several decades for physicists and astronomers to accept that black holes could even exist. Were they now going to have to come to terms with the idea that they were not completely black?

There were two surprising things about Hawking's results. Not only had he demonstrated that a black hole's darkness

HAWKING'S LAST JOURNEY

cannot exist forever, he had done so while his body was slowly ceasing to function. Hawking suffered from a neurological disorder called amyotrophic lateral sclerosis, also known as ALS. He had made his striking discovery in the face of incredible difficulty.

Stephen Hawking was born on 8 January 1942.[2] He grew up in the small cathedral city of St Albans, just north of London. He was a talented but headstrong schoolboy, who listened to Mozart, built one of the school's first computers and had a philosophy discussion club with his friends. He was a fast learner, but was rather lazy and was known for his blasé attitude to the school's difficult theoretical exams.

Hawking started studying physics and maths at Oxford University. While there, he felt the first indications that something was not right with his body. His speech would occasionally slur and he had difficulty tying his shoelaces. From time to time, he fell over. Hawking ignored the symptoms, but when he returned to his family home for Christmas in 1962, his parents arranged for a doctor to examine him.

After two weeks of investigations, he was given the devastating news that the nerves governing his motor functions were gradually atrophying. Hawking's brain would continue to function, but his condition would deteriorate until he became unable to breathe. The doctors gave him two years to live.

What do you do when you find out your life is about to end? Hawking, just twenty-one years old, isolated himself and listened to Wagner. But he didn't give up. While in hospital, he'd seen a boy die of leukaemia, and had the realization that others had it worse than him. What's more, he'd fallen in love with another student named Jane Wilde, who was studying modern languages in London. She gave him back his lust for life. They became a couple and started to dream of getting married.

Hawking continued his studies, becoming a PhD student at the Department of Applied Mathematics and Theoretical Physics at the University of Cambridge. He was trying to answer two of the biggest questions physicists can ask: what happened at the Big Bang? And what happens deep inside a black hole? He was inspired by the British mathematician Roger Penrose's analyses of black hole singularities, and started using the same methods to explore the singularity that existed at the moment of the Big Bang. This previously rather lazy student suddenly threw himself into his work. His investigations proved fruitful, respect for him grew among other physicists, and, most significantly, he had survived longer than the two years the doctors had told him he could expect to live. But his legs often collapsed under him and he had to walk with a stick.

THE MEETING IN MOSCOW

In 1973 Hawking travelled to the Soviet Union. By then, his stick had been replaced by a wheelchair and his voice had become slurred and difficult to decipher. He was accompanied by his wife Jane and the US physicist Kip Thorne.[3]

At this time, the Soviet Union was a superpower with a nuclear weapons arsenal to rival that of the United States. The Soviet Union's leader Leonid Brezhnev was holding talks with the US president Richard Nixon, aimed at reducing the risk of nuclear war.

At a hotel near Red Square, Hawking met the man behind the Soviet Union's nuclear weapons programme, one of the country's foremost physicists, Yakov Zeldovich. In the sixties Zeldovich had begun to analyse a number of cosmic phenomena, from black holes to the evolution of the universe. He was one of the first to

HAWKING'S LAST JOURNEY

propose that black holes might be detectable through X-rays, and that they were the engine driving the light from quasars.

In the hotel room in Moscow, an energetic Zeldovich told Hawking that he had discovered a peculiar characteristic of black holes: they can create particles from the void. Zeldovich had arrived at this strange result with his student Alexei Starobinsky. They had explored what happens in the vicinity of a black hole that is rotating so fast the space around it starts to come to life. To make sense of it, Zeldovich and Starobinsky applied two of the twentieth century's most important theories about the fundamental properties of nature: general relativity and quantum field theory. Einstein's great insight with the general theory of relativity was that gravity arises from the curvature of space and time. But it's a theory that describes the world in its big, heavy form: asteroids, planets, stars and galaxies. For the small, light world of atoms and subatomic particles, something else was needed, and that something was quantum field theory. This is a theory that builds on quantum mechanics and describes how electrons, photons and other subatomic particles can be born, morph into one another – and vanish.

There was one aspect of the quantum mechanical world that particularly appealed to visionary physicists like Zeldovich: the characteristics of the vacuum. An atom consists mostly of emptiness. Its nucleus is as small in relation to its whole as a piece of gravel is to a football stadium. But, paradoxically, the subatomic vacuum is not completely empty. Inside it, virtual particles are constantly coming into being and ceasing to exist – and they do so in pairs. This is because when it comes to energy in nature, the books must always balance. If one particle has positive energy, its pair must be negative, making their total energy equal to zero. When these particles are born in a vacuum, they dash off in separate directions, before being reunited and vanishing. No energy has been created or lost during their mayfly lives. The particles bear the epithet 'virtual' because they lack the material world's solidity and permanence. The quantum mechanical void is full of them, but they are as intangible as sunlight glittering on the surface of the sea.

Zeldovich and Starobinsky had calculated that virtual particles close to a rotating black hole can acquire enough energy from the spinning spacetime to make the leap from the virtual world to the real one. It might appear that the black hole is creating particles from nothing, but in reality it is the black hole's rotational energy that provides the means for this transformation.[4] As a result, the rotation slows over time. Zeldovich and Starobinsky assumed that when the black hole finally stopped rotating, the release of particles would come to an end too.

Hawking was fascinated by Zeldovich and Starobinsky's idea, but remained sceptical. Their theory was premised on a mix of elements from general relativity and quantum field theory. No one knew for certain what such a combination would lead to. When Hawking returned to Cambridge, he couldn't stop thinking about their radical suggestion. He wanted to carry out the calculations for himself. For several months, over the autumn and winter of 1973, he tried to find out whether black holes really can glow by creating particles out of the vacuum. He went even further than Zeldovich and Starobinsky, investigating whether it was possible for black holes to continue creating particles even after they had stopped rotating.

BLACK HOLES ARE NOT BLACK

Because of his neurological condition, Hawking couldn't do his calculations with pen and paper. Colleagues and students

HAWKING'S LAST JOURNEY

helped by taking notes, and he even carried out many of the long, difficult calculations in his head. In the end, Hawking concluded that Zeldovich and Starobinsky were both right and wrong. It was true that black holes could shine, but Hawking demonstrated that all kinds of black holes, not only rotating ones, emit radiation.

In his popular science writing, Hawking has described his discovery by following what happens to a pair of virtual particles close to the event horizon.[5] Generally, these particles will, as mentioned above, be created, move apart, reunite and then finally disappear. But close to the black hole they can be torn apart. One particle travels into the black hole while the other moves away from it. The reunification never takes place, and the particles step across the threshold to reality and acquire a material existence.

The constant morphing of virtual particles into real ones means that a swift stream of particles appears to flow from the black hole.[6] Hawking demonstrated that this stream consisted primarily of nature's lightest particles, photons (that is, light) and neutrinos. He also showed that the energy of the particles is perfectly described by the formula for heat radiation that physicists are so well acquainted with here on Earth.

Hawking was taken aback by his own discovery. He had shown that black holes are not black. But that wasn't the only consequence of this revelation. As a black hole radiates, it loses energy and its mass decreases. Hawking had shown that the lower a black hole's mass, the more it would glow. The radiation leads to a self-perpetuating process that ends in an inferno around the black hole, from which all the various kinds of particles nature is capable of producing are spewed out. Ultimately, the black hole explodes and disappears. What was once the universe's darkest object has vanished in a cascade of radiation.

Hawking calculated the time it would take for a black hole to radiate away all its mass and vanish. Using the formula for how much a black hole of a given mass radiated, Hawking was able to calculate its temperature. The larger the mass, the lower the temperature. A black hole with the same mass as our Sun would have an incredibly low temperature: only a millionth of a degree above absolute zero ($-273.15°C$). This black hole would produce so little radiation its mass would decrease extremely slowly, giving it a lifetime of 10^{67} years (1 followed by 67 zeros), which is 10^{57} times longer than the current age of the universe. For a supermassive black hole with a mass equivalent to several million Suns the temperature would be even lower: 10^{-15} degrees (0.000000000000001) above absolute zero. Its lifetime would be unimaginably long: 10^{94} years. This means that, of all the objects in the universe, the largest black holes will be the longest lived.

But the radiation from these black holes is so weak it's impossible to observe. All the black holes we know of have a temperature lower than that of the space surrounding them. The universe is full of the cosmic background radiation discussed earlier, which has a temperature of 2.7 degrees above absolute zero. Since heat flows from high to low temperatures, these astrophysical black holes will absorb radiation from their surroundings, rather than sending out any of their own. Even if we were able to place a telescope near them, we would not be able to observe empirically the radiation Hawking's theory predicted.

But Hawking calculated that a black hole formed in the universe's infancy and having a mass of one trillion kilograms (equivalent to a small mountain on Earth) would have been radiating so long it could disappear in our current cosmic epoch. In an article entitled 'Black Hole Explosions?' he wrote that this kind of black hole would blast apart with the force of a million hydrogen bombs.[7] This was the takeaway from the article whose contents Hawking presented that spring of 1974, when ABBA sang about Waterloo.

When astronomers searched for that radiation – later dubbed Hawking radiation – it turned out to have an unexpected consequence for us on Earth.

HAWKING RADIATION AND THE INTERNET

At the end of the seventies the Australian radio astronomer John O'Sullivan became fascinated with Hawking's theory. Hawking's colleague Martin Rees had suggested that *if* black holes did explode in the way Hawking had predicted, the explosion could lead to a short burst of radio signals. In an article from 1977 Rees wrote that signals created 'by a single entity of subnuclear size' would be possible to see 'from as far away as the Andromeda galaxy!'[8]

O'Sullivan decided to search for these radio signals.[9] He and his colleagues at a radio telescope in Dwingeloo in the Netherlands developed new techniques to quickly scan and analyse the sky's radio noise. In the end, they didn't find any radio signals from exploding black holes, but the algorithms they established found an important application in another arena.

O'Sullivan moved back to Australia and started working for CSIRO Division of Radiophysics, outside Sydney. There, he tried to come up with a way of sending data over wireless networks within offices or in lecture halls. The wireless signals reverberated around the room and got mixed up, so picking up a clear signal was difficult. But O'Sullivan realized that the algorithms developed in the hunt for Hawking radiation could be utilized for this new wireless technology, which was known as WLAN (Wireless Local Area Network). 'The problems we had to solve in radio astronomy back then with black holes and later with WLAN were remarkably similar,' O'Sullivan reflected later in life.[10] The network protocol he and his colleagues created to send information wirelessly between different computers is today called IEEE 802.11, or, more commonly, Wi-Fi.

Every day, this wireless network protocol connects people and machines all over the world. That the technology has its roots in the hunt for signals from black holes is not without a certain irony. Someone who got stuck *inside* a black hole would lose the ability to communicate with the outside, but the hunt for exploding black holes made it possible for a huge proportion of the people on Earth to communicate with one another.

Blue-skies research – science pursued from pure curiosity about how nature works, and without any practical application – is one of the best ways to drive technological innovation. O'Sullivan's search for Hawking radiation and the way it led to the development of Wi-Fi is an excellent example of this. But even if these technological applications are important, the greatest benefit of blue-skies research is still purely existential: we learn more about nature and our place within it. This kind of benefit cannot be quantified in financial terms.

A GLOBAL CELEBRITY IN THE MAKING

Hawking's realization that black holes are not completely black opened up a whole new branch of research, and his renown in the physics world grew. He won countless awards, was elected to the prestigious Royal Society in London, and became Lucasian Chair of Mathematics at Cambridge, the same professorship Isaac Newton had once held. He thrived as a researcher, and said that his disability had in effect given him an advantage, since his

HAWKING'S LAST JOURNEY

exemption from academic duties, such as teaching and committee work, gave him more time to think about physics.

But while Hawking's career as a physicist was flourishing, his body was withering away. In 1985 he got a serious infection in one lung that made it almost impossible for him to breathe. The doctors were forced to open his trachea and install a mechanical ventilator. Hawking was able to go on living and breathing, but he could no longer talk.

Paradoxically, though, the loss of his already weakened voice meant Hawking could communicate better than before. He was provided with a specially designed program to enable him to speak via a computer, using a control panel into which he could enter text. The computer then read out his sentences, with his Speech Plus CallText 5010 hardware synthesizing the robot voice that was to become Hawking's hallmark.[11]

As well as giving lectures to a growing audience, Hawking also wrote the popular science book *A Brief History of Time*. It has sold more than twenty-five million copies, and is still one of the world's bestselling popular science books. Hawking became a celebrity around the world, and an icon of theoretical physics. He was afforded the same status as Newton and Einstein, got to meet the pope, and turned up in both *The Simpsons* and *Star Trek*. His contributions as a science communicator were on a par with his theoretical ones.

But despite Hawking's successes, cracks were forming in his family life. '[Sometimes] life was just so dreadful, so physically and mentally exhausting, that I wanted to throw myself in the river,' his first wife Jane Hawking said.[12] She was looking after her husband, but found herself constantly in his shadow. The physical and emotional caretaking took a profound toll on her, as well as forcing her to put her own doctoral studies on hold.

The situation echoes that of Einstein's first wife, the physicist and mathematician Mileva Marić. She too was overshadowed by her spouse, and wrote, 'One gets the pearl, the other gets the box it came in.'[13]

Both Einstein and Hawking remarried, Einstein to his cousin Elsa Einstein, and Hawking to his nurse Elaine Mason. She'd previously been married to the engineer who'd constructed the computer that enabled Hawking to take his synthetic voice everywhere he went. Einstein and Hawking became the two best-known physicists of the twentieth century, but they were men who would never have achieved the success they did had it not been for the support, and sacrifices, of their wives.

THE INFORMATION PARADOX

Hawking's discovery that black holes can give off radiation lent a new urgency to the question of what happens to the information that falls into a black hole. Imagine your friend has just thrown a book into a black hole. You wait until the black hole has evaporated, leaving only its Hawking radiation, scattered through space. Your friend tells you the book contained one of Shakespeare's plays. Would it be possible for you to analyse the remaining Hawking radiation to find out whether it was *Othello*, *Hamlet* or *Romeo and Juliet* that had disappeared inside?

If your friend had thrown the book into a fire, it would *in principle* have been possible to reconstruct the content by carefully observing all the radiation, smoke and ash that remained after the book had burned up. Of course, it would be impossible in practice, but according to the known physical laws, all processes occurring after the book has been burned should be traceable in reverse, thus providing information about the

HAWKING'S LAST JOURNEY

material constitution of the book and its content. Would this also be true of the Hawking radiation left after the book has disappeared together with the black hole it was thrown into? At stake is the fundamental issue of whether information is retained in all the physical interactions that occur in nature. If you're unable to use the Hawking radiation to reconstruct which of Shakespeare's plays got thrown in the black hole, that means the information has been lost forever. It's not simply a question of the information having disappeared behind an event horizon. That would be OK – even if the book were merely locked in a cabinet, I would be unable to find out more about its content. The issue here is that the event horizon and the black hole would ultimately disappear completely. It's as though both the cabinet and the book might disappear without trace.

The idea of information being permanently erased prompts deep unease in physicists. A physical theory builds on the possibility of knowing what will happen in the future and what has happened in the past. It should be possible to use information about what is happening *now* to describe what will happen *tomorrow* and what happened *yesterday*. Perhaps it's not always possible in practice, but it should at least be possible in principle. Nature should be predictable. But if information can disappear forever, it becomes impossible to use the laws of nature to reconstruct what has happened in the past and what will happen in the future. The implication of this is that nature contains a fundamental unpredictability.[14]

The question of whether black holes could completely erase information has given rise to a problem known as the *information paradox*. Quantum mechanics tells us that information is always retained. Therefore, it shouldn't be lost when a black hole disappears leaving nothing but its Hawking radiation. But the general theory of relativity tells us that the book containing

Shakespeare's play will pass through the event horizon without anything particular happening to it. The information in the book ends up in the black hole, and there is no possibility of it getting transferred to the Hawking radiation on its passage through the event horizon. After the black hole has evaporated, the information is gone.

These two theories seem to give different answers to the question of what happens to the information that falls into a black hole. To resolve the information paradox, the two theories must be combined in a non-contradictory way, leading to a new understanding of the fundamental laws of nature. Black holes have therefore proved to be fertile objects of study when exploring how different subdisciplines of physics can be brought together under a common theoretical framework. What physicists have come to understand is that, in order for this convergence to occur, lessons will have to be taken from the branch of physics known as thermodynamics.

THE THERMODYNAMICS OF BLACK HOLES

Thermodynamics is the study of the relationship between heat, work, temperature and energy in a system. These relationships were the subject of much debate among nineteenth-century physicists and engineers in the wake of the industrial revolution. They were led by practical concerns, such as wanting to find out how much energy was stored in coal and water power, or ensuring that locomotives and steam-driven machines functioned as effectively as possible. It was Jacob Bekenstein from the University of Texas at Austin who realized, in the early seventies, that analysis of these practical phenomena would provide new insights into the inner workings of black holes. He discovered that black holes challenged physicists' understanding of one of the universe's most important principles: the second law of thermodynamics.

The first law of thermodynamics states that the total energy of an isolated system is always constant. Energy can change form and turn up as heat, electricity, nuclear bonds, movement and so on, but the amount of energy does not change. The second law concerns a term that has a slightly mystical aura, but is actually very straightforward: entropy.

Entropy can be illustrated using our day-to-day experiences. If we can't be bothered to tidy a room, it gets messy. If we don't keep the contents of our hard drive in check, it quickly turns into a jumble of folders and files (something I've unfortunately experienced while writing this book). Entropy is about the tendency of things to become more disorderly over time. The second law thus states that, in an isolated system, entropy will never decrease. Either entropy is constant, or, more commonly, it increases.

In 1870 the Austrian physicist Ludwig Boltzmann introduced a new way of looking at entropy that would, a century later, become central to our understanding of black holes. Boltzmann realized that entropy was about statistics. There are more ways for a system to be disordered than ordered, hence the disorder growing over time. Let's think back to that room and imagine we eventually get round to tidying it. You put all the objects where they belong – the pen in the pen holder, the plant pot on the windowsill, the pillow on the bed and so on. When the room is tidy, it is ordered. But then you take a pen and put it in the wrong place. There are many wrong places: on the bed, in the plant pot, on the floor. When the pen is in the wrong place, the entropy has grown a little. Now take the plant pot and put that too in the wrong place. The entropy has grown even more, because the number of places in which both the pen and the pot can be out of place is greater than for just the pen.

If all the objects in the room are in the wrong place, you have reached maximum disorder. At that point, the room is at its highest level of entropy. But there are several ways this maximum disorder can be arrived at, since the pen, the plant pot and the pillow can be scattered in many incorrect locations. Boltzmann's realization was that entropy is constantly increasing because the number of possible disordered states is larger than the number of possible ordered states. That's why a system is more likely to move towards a more disordered state than an ordered one. The only way to counteract this statistical tendency towards disorder is to provide a system with energy from the outside – which, in our case, means tidying the room.

Boltzmann's statistical view of entropy was central to Jacob Bekenstein's argument about the entropy of black holes. Imagine a star falling into a black hole. The star has a certain degree of entropy. When it has entered the black hole, it has vanished for us on the outside. Its entropy is no longer part of the outside universe, so it seems like the entire universe's entropy has decreased. The second law of thermodynamics has been broken!

In order to adhere to the principle that entropy must always increase, Bekenstein suggested that a black hole must also have entropy. When the star enters the black hole, the black hole's entropy increases by an amount equivalent to that of the star. But how? It is possible to conceive of the entropy of a star in terms of its physical properties, such as the number of particles, its temperature and other variables. But it was less clear how the spacetime geometry of a black hole was related to entropy. Hawking had previously shown that, according to general relativity, a black hole's surface, that is, the area of its event horizon, could only grow when the black hole devours the star. Bekenstein therefore wondered whether this too might be related to the way entropy increases. Via an intricate thought experiment, he came to a far-reaching conclusion: the entropy of a black hole must be proportional to its surface area.

Hawking was surprised and annoyed by Bekenstein's suggestion, and initially dismissed it. Black holes were described using their mass and spin rate (and electrical charge – though, as we saw in Chapter 6, that can be overlooked for astrophysical black holes, which are electrically neutral). Entropy, on the other hand, was a thermodynamic quantity, which, at first glance, could have nothing to do with black holes. But once Hawking had discovered that they had a temperature and that they glowed, he realized that Bekenstein was onto something important. Hawking calculated the entropy of a black hole and it turned out to be enormous. A black hole with a mass equivalent to that of the Sun would have an entropy significantly bigger than that of all the particles in the universe put together. Black holes have the greatest entropy of any known object! This was surprising, because it showed that Einstein's general theory of relativity was not enough; to understand black holes we need more. General relativity says that black holes are the simplest objects in the universe, since they are described only by their mass, spin rate and their (overlookable – see above) electric charge. But what Hawking, Bekenstein and others realized was that quantum mechanical effects close to the event horizon presented a new image of the properties of a black hole. From having been one of the very simplest objects, they now seemed, instead, to be one of the universe's most complex.

The conclusion, therefore, was that the spacetime description of black holes does not tell us everything we need to know about them, because it excludes their thermodynamic properties. But *what* exactly is it in a black hole that has entropy? Boltzmann had realized that entropy was a measure of the amount of disorder in a system. The various entities in that system – like the particles in a gas or the zeros and ones on a hard disk – could be arranged in various ways, and the most disordered states would have the greatest number of possible configurations. Several physicists have therefore explored the possibility that black holes may consist of a great number of small parts, for example the tiny, vibrating strings that string theory says form the fundamental entities of nature. It's these microscopic parts of a black hole that give rise to its entropy.

But entropy can also be seen as a measure of hidden information. Let's return to the untidy room. Imagine it's been thoroughly tidied and everything is in the right place. At that point, it has low entropy. If I ask where the pen is, you can reply, without looking, that it is in the pen holder, because you know it's where it should be. But if the room is untidy and its entropy is high, the pen could be in numerous other places instead – on the window, for instance, or under the bed or behind the sofa. When I ask you where the pen is, you are unable to give a definite answer. Information about the pen's location is hidden from you.

In a comparable way, the entropy of a black hole may have something to do with our lack of knowledge about what formed that black hole. From the outside we have no way of knowing what has fallen into the black hole. It might have been planets, stars, other black holes, astronauts – anything. In this sense, entropy is a measure of all the different ways through which the black hole could have come into being. Which of them has actually occurred will always remain hidden to us.

THE HOLOGRAPHIC PRINCIPLE

The entropy of a black hole has one strange property that differentiates it from other systems. The gas in a container has an entropy proportional to the volume of the container. But the entropy of a black hole is proportional to its surface area, not its volume. If the entropy is related to the hidden information about all the ways the black hole could have come into existence, it seems that this information is retained in two dimensions, rather than three.

This result has led to one of the most exciting developments in modern theoretical physics, known as the *holographic principle*. It's an idea that has been developed by several physicists, such as Leonard Susskind from the US, the Dutch Nobel laureate Gerard 't Hooft and the Argentinian Juan Maldacena.[15]

According to the holographic principle, all the information in nature exists in a space that is one dimension lower than we experience it. This might seem strange: if we think of a hard disk, for example, we know it's a three-dimensional object that has a certain volume. The information stored on it is therefore part of this three-dimensional, physical object. But just as a hologram is stored on a flat surface but creates a three-dimensional projection, supporters of the holographic principle think that information about our world exists in a kind of fundamental, two-dimensional space and that the three dimensions we see around us are a projection from this space. It sounds abstract – and it is. But the physicist Maldacena has demonstrated that there is an underlying similarity between a particular kind of quantum field theory in a certain number of dimensions and general relativity in a higher dimension. The entropy of black holes has thus inspired scientists to take their theoretical investigations along new paths and think in completely new ways about how

280 FACING INFINITY

information is stored in nature. Black holes are found not only at the centre of most galaxies; they are also at the centre of contemporary developments in theoretical physics.[16]

Over the years, several solutions to the information paradox have been put forward.[17] One of the weirdest is called black hole complementarity.[18] According to this, both of the assertions of the information paradox are simultaneously true. An outside observer could extract all the information that had been encoded in the Hawking radiation close to the event horizon, but someone who fell into a black hole would notice that the information had gone inside and would never be able to exit. These two contradictory statements are simultaneously true because the same observer cannot confirm both truths at the same time. It's as though, in the case of the black hole, reality gets divided into two versions that can never overlap. A consequence of this theory is that there could be a 'firewall' in the vicinity of the event horizon, and that this firewall quickly annihilates every observer who enters a black hole.

The thinking around black holes is speculative and can be difficult to follow. Unfortunately, we do not yet have any experiments that prove that Hawking radiation exists or that nature works in the ways supposed by the adherents of the holographic principle. But, for a brief period in 2008, it seemed like Hawking radiation might have a decisive role to play on Earth.

BLACK HOLES ON TRIAL

'Are we all going to die next Wednesday?' screamed the headline of an article in the *Daily Mail* on 4 September 2008.[19] The newspaper described a doomsday scenario: earthquakes rocking the Earth's surface, coastal areas drowned by tsunamis, lava spewing

HAWKING'S LAST JOURNEY   **281**

everywhere. In the end, Earth would collapse in on itself completely. A black hole would have swallowed our planet.

The *Daily Mail* painted this apocalyptic picture because the particle accelerator the Large Hadron Collider (LHC) in Switzerland was about to go into operation. In the accelerator, protons would collide at high speeds, turning their energy into a shower of other particles in the process. By observing how often these transformations happened and what they led to, physicists hoped to find out more about the fundamental properties at play in the world of particles. Because the energy in the collisions was equal to what particles may have had during the first, tumultuous moments of the universe, learning more about these characteristics would also provide information about the conditions that prevailed in the early universe. One of the foremost objectives was detecting the Higgs Boson, the last piece of the puzzle in physicists' standard model of particle interactions.

But some people were critical of the entire operation, afraid that the high-energy collisions would produce microscopic black holes that would grow in size, eventually swallowing the world from within. Some of these critics sued CERN, the organization that runs the LHC, at the European Court of Human Rights, seeking to prevent the experiment going ahead. In Hawai'i too, a lawsuit was filed that would require all funding for CERN to be withdrawn. Members of Romania's Conservative Party demonstrated outside the offices of the European Commission in Bucharest to register their concern that the LHC would create these black holes, and a spokesperson for the party suggested that the particle accelerator risked 'the safety of our planet'.[20]

The critics' fears were not unjustified. In 2002 two scientists had argued that microscopic black holes could be created in the particle accelerator, and the idea was discussed in scientific circles.[21] A website published by the well-regarded Max Planck

282   FACING INFINITY

Institute of Gravitational Physics in Germany ran an article entitled 'Particle Accelerators as Black Hole Factories?'

The LHC can accelerate particles to kinetic energies (that is, the energy associated with the motion of the particles) equivalent to two mosquitoes flying into each other. That may not sound like much, but when billions and billions of particles are zooming towards each other in the LHC's tunnels, it adds up to the same energy as a salvo of cannon fire from a warship. This energy is not enough to create a black hole, but in the 2002 article the scientists had argued that the laws of nature could work differently in the miniature world of particles. Perhaps, at this infinitesimal scale, the strength of the gravitational force would increase, making it possible for microscopic black holes to form in the collisions.

No physicist believed that a black hole created in this way would grow to a size at which it could swallow the Earth, but among those who wanted to stop the experiment, this was the prevailing view. The situation wasn't helped by the prominent physicist Nima Arkani-Hamed telling the *New York Times* that 'the Large Hadron Collider might make dragons that might eat us up'.[22]

To assuage the critics' fears, CERN set up a review that aimed to confirm there was no risk of microscopic black holes eating up the Earth. The assessment group did not deny that such black holes might be formed in the particle collisions, but found that their Hawking radiation would lead to them evaporating rapidly. 'Any microscopic black holes produced at the LHC are expected to decay by Hawking radiation before they reach the detector walls,' the authors of the report reasoned.[23] But the critics asserted that, since no one had observed this radiation, there was no certainty it actually existed. The CERN report had, however, provided an additional argument: *if* it were possible for microscopic black holes to be created and to swallow the Earth, it would have happened a long time ago. The Sun, the Moon, the Earth and all the celestial bodies are being bombarded every second by cosmic radiation. This radiation consists of particles that come from our own galaxy and others. They can have enormous energy, a sign of their violent origins: exploding supernovas, intensely magnetic neutron stars and, indeed, the environments around black holes. The energy of these cosmic particles is greater than that of the particles colliding in the LHC. The assessors calculated that in the universe, 10,000 billion of these collisions happen *every second.*

Imagine now that each of these collisions created a microscopic black hole that would grow and ultimately swallow the object the particle collided with. If that happened, we would see it every time we looked up at the sky – and on the Earth too. The chief argument against black holes being created in the LHC and gobbling up the Earth is that if it were possible for this to happen, it would have done so long ago as a result of those cosmic particles.

On 10 September 2008 the LHC was put into operation. Neither the European nor the US courts had gone along with the critics and stopped the experiments. In the underground tunnels, particles were flung around at close to the speed of light. But no Hawking radiation from microscopic black holes was detected – and neither were the earthquakes, tsunamis and volcanoes the *Daily Mail* had warned of. The Earth survived. In 2012, thanks to the LHC, physicists observed the particle they'd been hoping to see: the Higgs Boson. Instead of black holes and the end of the world, the particle accelerator led to a Nobel Prize in Physics for Peter Higgs and François Englert, who'd predicted the existence of the Higgs Boson. A few years later, Hawking jokingly commented on the fact that the accelerator had not produced a hint of radiation from black holes, saying, 'This is a pity, because if they had I would have got a Nobel Prize.'[24]

Although microscopic black holes were not seen at the LHC, physicists have nevertheless come up with several other scenarios for how minute black holes could have made an impact on our planet. On 30 June 1908 there was a large explosion near the Tunguska river in Eastern Siberia in Russia, with its shock wave levelling some eighty million trees across a huge area of forest. It was the largest cosmic impact in recorded history, and had it hit densely populated areas, mass death would have followed in its wake. The explosion was most likely caused by a comet or asteroid impact in the atmosphere, but there was one very mysterious aspect to it: there was no trace of a crater. In 1973 two scientists from the University of Texas at Austin published an article in *Nature*, where they speculated that the lack of an impact site could be explained if the explosion were caused by the collision of a microscopic black hole with the atmosphere.[25] The scientists made direct reference to a recent calculation by Hawking that showed how such primordial black holes could have been created very early on in the universe and exist in large numbers in the cosmos today. We encountered them in Chapter 9, where they appeared as seeds that could have grown into later day supermassive black holes. In general, such primordial black holes could have a wide range of possible masses. The Tunguska black hole would have had a mass equivalent to that of a large asteroid and a size as small as an atom. Being so small, it would not create a crater but instead produce a shock wave in the air before continuing straight through the Earth, exiting some-where in the North Atlantic. The scientists behind the idea even suggested that maritime records should be checked for traces of the exit, which would prove the black hole-impact hypothesis. The historical archives were indeed investigated, but no evidence was found.[26]

Today, the idea that a black hole caused the Tunguska event is not taken seriously, but it does show how physicists entertain the possibility of black hole impacts on events here on Earth. One idea holds that primordial black holes whizzing through our planet could leave geological imprints in ancient boulders, in the form of tiny elongated tubes.[27] If found, the tubes could be proof of the primordial black holes that Hawking hypothesized would exist. Such primordial black holes could damage human tissue if passing through our bodies. A recent calculation showed that black holes with a mass less than that of a small asteroid would pass through our bodies unnoticed (basically being too small and passing too quickly to break up human tissue).[28] For larger masses, however, an encounter of this kind would end badly. Luckily, even if the hypothetical primordial black holes did exist, such interactions would be so infrequent that your chance of getting killed by a black hole that has been speeding through the universe since just after the Big Bang is extremely low.

Yet another speculation proposed by scientists is that a primordial black hole could exist on the outskirts of our solar system.[29] Astronomers have suspected for some time that an as-yet-undetected planet – dubbed Planet Nine – exists somewhere far beyond the orbit of Pluto. In 2019 a team of researchers suggested that Planet Nine was perhaps not a planet after all, but a primordial black hole. Since Planet Nine has an estimated mass five to ten times that of the Earth, such a black hole would be roughly the size of an apple. The existence of Planet Nine itself is still an open question, so to conceive of it as a black hole adds yet another layer of speculation. Nevertheless, the prospect of a primordial black hole orbiting the Sun at vast distances shows how far the ideas that Hawking helped develop have led scientists in their conceptions of what is possible.

HAWKING'S LAST JOURNEY

On 14 March 2018 Stephen Hawking went to sleep for the final time at his home in Cambridge. His daughter Lucy was at his side. It was snowing, and shortly before her father died, she went out into the family's garden and built a snowman. She turned its head to the sky, like an astronomer contemplating the secrets of the universe. Hawking had demonstrated that black holes cannot exist forever, but he had also identified a gap in our understanding of them that still keeps a large number of scientists busy today.

Hawking was closely associated with the city of Cambridge, where he spent the majority of his academic career. Thousands of mourners followed his funeral car, which was filled with roses, tulips and lilies in yellow, red and white, as it made its way through the city. When the hearse stopped so that the black-clad pallbearers could carry the coffin into Great St Mary's Church, the crowd that had gathered to bid farewell broke into a spontaneous round of applause. The church bell rang out seventy-six times, one for each year Stephen Hawking had lived.

When he was diagnosed with ALS, the doctors had given him only two years to live, but he survived more than half a century, becoming a successful physicist, the father of three children and an international superstar. Hawking had carefully cultivated his image as a theoretical physicist in the public eye, and people around the world had seen him as a source of inspiration. Behind that image and the extraordinary feats of his intellect was, however, an all-too human being. He could be stubborn and arrogant, a trait that extended towards friends and colleagues. 'I was nervously aware that his arrogance was in poor taste and was putting me in danger of losing me my friends, if not my relations,' Jane Hawking wrote in her autobiography. That arrogance sometimes turned into impishness, such as his reported liberal use of the

HAWKING'S LAST JOURNEY

wheelchair to run over people he didn't like. A rumour had it that Hawking regretted not having run over Margaret Thatcher's toes, which he denied, adding, 'I'll run over anyone who repeats it.'[30]

As Hawking's stardom grew, his fame and prestige led him to rub shoulders with the ultra-rich, such as visiting Jeffrey Epstein's private Caribbean island during a conference on gravity funded by Epstein (held before the former financier was arrested and eventually exposed for sexual trafficking of minors), and publicly partnering with billionaire Yuri Milner and Facebook founder Mark Zuckerberg on a project to try to send a miniature spacecraft to Alpha Centauri.

'His celebrity status gives him instant credibility that others do not have,' complained Nobel Laureate Peter Higgs. This credibility often led his remarks on topics far from his own research area to gain widespread attention (such as his doom-laden prophecy that '[t]he development of full artificial intelligence could spell the end of the human race').[31] Frank Wilczek, who was one of Hawking's colleagues and also a Nobel Laureate, reflected, 'His status as an idol was also, I think, hard on Stephen. He knew better. Less would have been more.'[32] To idolize Hawking shrouds the human struggle both he, and those around him, had to endure as he broke new ground as a physicist and a communicator of science.[33]

At the end of his life, Hawking spoke less and less frequently. Writing messages for the robot voice took longer and longer, and his phrases grew ever shorter. Even though it got harder for Hawking to communicate, he continued to collaborate with colleagues to the very end, trying to solve the information paradox of black holes.

Hawking left behind a final message. On 15 June 2018 Nobel Prize laureates and bishops, actors and students, scientists, politicians, musicians and Hawking's family gathered in Westminster

Abbey to say their farewells. Two bishops lowered Hawking's ashes into a grave in the stone floor of the church, not far from the last resting places of Isaac Newton and Charles Darwin. 'Here lies what was mortal of Stephen Hawking', it says on the grave, alongside an engraving of a black hole and Hawking's formula for the radiation that bears his name.[34]

At the same time as Hawking was being buried in Westminster Abbey, a several-hundred-tonne antenna outside the village of Cebreros in Spain was slowly rotating.[35] Ordinarily, the European Space Agency used the antenna to communicate with spacecraft and satellites, but now they were making it point in the direction of V616 Monocerotis.[36] This system consists of a star and a black hole and lies about 3,500 light years away in the constellation Monoceros. At the time it was the closest-known black hole to Earth (as mentioned in Chapter 6, the closest observed today is Gaia BH1 at 1,560 light years away).

A current flowed from the electronic insides of the antenna and surged out of the metal colossus in the form of radio waves. The signal shot out into the Earth's atmosphere and on into space. The radio waves carried a message Hawking had recorded before his death. The Greek composer Vangelis was commissioned to put the message to music with his ambient synth textures, and the recording was played both at Hawking's funeral at Westminster Abbey and in the radio waves shooting out through the cosmos. 'I am very aware of the preciousness of time,' Hawking said. 'Seize the moment. Act now.'

The European Space Agency wanted to honour Hawking's memory by sending his final words to the nearest-known black hole. 'I have spent my life travelling across the universe inside my mind,' Hawking's message continued. 'Through theoretical physics I have sought to answer some of the great questions.' Hawking had posed a question physicists are still trying to answer:

if something falls into a black hole, is it really gone forever? Perhaps the question can also make us value what is important in our lives, since black holes warn us of the risk that everything can disappear.

Hawking lost his physical voice and his mobility, but thanks to his family, friends and colleagues, he was able to explore the secrets of space. 'I have been enormously privileged through my work to be able to contribute to our understanding of the universe,' he said in the message that is still zooming through space at the speed of light. And he added, 'But it would be an empty universe indeed, if it were not for the people I love and who love me.'

Hawking's words and Vangelis's music are the first signal humans have intentionally sent to a black hole. At some point in the middle of the sixth millennium, this signal will reach the black hole in V616 Monocerotis. Hawking lost his ability to speak, but his message will continue beyond the black hole's event horizon, facing infinity as no other human endeavour has done before.

Chapter 14

· · · · · · · · · · · · · ·

ARE WE LIVING
IN A BLACK HOLE?

We ran forward to attack. Bullets rained down. It didn't matter – we had been given a good slug of brandy to drink. We ran on. We battled on. In one bound the blockhouse was surrounded. It was just a smoking heap of stone full of horrifyingly mutilated Kraut corpses. We broke through. The Kraut fled. Suddenly I found myself face to face with a German officer who was aiming his revolver at me. Dispassionately, I bayoneted him through the chest, crack! I did it with so much force that the barrel went in too. I couldn't get it out. I grabbed the rifle of a dead soldier and ran on.[1]

On 6 April 1915 French soldiers attacked German posts on the top of Hartmannswillerkopf in the Vosges mountains. Just over a century later, I've come to visit the mountain. I've driven past picturesque vineyards, farms and villages in north-eastern France, and carried on up the winding roads of the mountain range. At the peak of Hartmannswillerkopf is a museum, and it's there that I read the soldier André Larrue's hair-raising letter about the attack.

It was from this mountaintop that the rumble of war rolled out across the landscape to where Karl Schwarzschild was working on his formula for black holes. 'My calculations on Einstein's

ARE WE LIVING IN A BLACK HOLE? **291**

gravity theory have developed well,' Schwarzschild wrote to his wife Else on Christmas Day 1915, adding, 'But the Frenchmen have destroyed our people's festive peace with their explosions up there on Hartmannswillerkopf, and yesterday a shell landed by a village churchyard, close to a funeral procession.'

French and German troops fought for years for control of the mountain. Around thirty thousand people sacrificed their lives. On a plaque in the museum I read about the field and foot artillerymen, the pioneers, the minesweepers, the infantry, the hunters, the machine-gun handlers, the cavalry, the communication troops, the medical staff and the armament troops, all highly specialized in their military tasks but completely unprepared to fight in the freezing cold and snow on this stony terrain.

I've travelled to the place the Germans call the 'Mountain of Death' to find out more about the exact conditions under which Schwarzschild discovered his now-famous formula. I have a sense that its origins are important. On the top of the mountain, in addition to the museum, there is a necropolis, a cemetery and the battlefield itself. Trenches, barbed wire, lookouts and fortifications have been preserved. Between the network of trenches lies the no man's land that separated the two armies.

I go down into a trench and gaze out through the barbed wire over the fertile slopes around the Rhine. Some of the trenches are linked by tunnels that lead into one part of the mountain and out of another. Inside these, the soldiers built traps for opponents chancing an attack. Heading into the tunnels meant putting themselves in mortal danger.

The Schwarzschild metric indicates that something analogous to these underground tunnels could also exist in spacetime. They turned up for the first time in mathematical form, just as black holes did. The same year Schwarzschild published his formula, the Austrian physicist Ludwig Flamm noted that it had a peculiar

292 FACING INFINITY

feature. The formula seemed to contain a kind of doubling of the world: it described ours, but it also described another world with similar geometric properties. For Flamm this was a mathematical curiosity, but twenty years later, Einstein demonstrated that the two worlds weren't so isolated from each other. Working with his colleague Nathan Rosen, Einstein calculated that the event horizon of a black hole connected the two worlds in what would later become known as an Einstein–Rosen bridge.[2]

Einstein and Rosen performed the calculations in order to explore whether nature's smallest particles, such as the electron, could be described using these peculiar spacetime geometries. Travelling over an Einstein–Rosen bridge from one world to the other is impossible, however – if you wanted to do so, you'd have to travel at faster than the speed of light. But the mathematics roused the physicists' imaginations. Perhaps there were more examples of such passages between the two parts of spacetime?

In 1957 the US physicist John Archibald Wheeler – the one who spread the term 'black hole' – named these spacetime passages 'wormholes', taking inspiration from the channels maggots burrow from one part of an apple's surface to another.

Wormholes have become a staple of science fiction. In *Interstellar*, one has opened up just by Saturn. The astronauts who enter the wormhole emerge in another galaxy, close to the gigantic black hole Gargantua. The thinking is that a wormhole has two exits. They can be in different parts of the universe, or even connect different universes. The openings could be many millions of light years apart, but the passage through the wormhole might reduce the distance substantially, creating a shortcut through space.

But Wheeler and a colleague demonstrated that the wormholes Einstein and Rosen had described were incredibly unstable, rather like a bridge built of soap bubbles. Such a bridge would

ARE WE LIVING IN A BLACK HOLE?

exist for just a fraction of a second before the bubbles burst and the bridge disappeared. Similarly, a wormhole – should one ever exist – would vanish so quickly neither matter nor light could pass through it.

Wheeler's student Kip Thorne (who was a scientific advisor on the production of *Interstellar*) continued to analyse wormholes through the complicated mathematics of the general theory of relativity.[3] He demonstrated that a wormhole could be stabilized for short periods, but that this would require a strange form of energy produced through quantum mechanical processes. These processes take place at such a microscopic level that it's hard to envisage a wormhole big enough for astronauts, asteroids or other large objects to pass through. I asked Kip Thorne what his thoughts were on wormholes today, and he answered, 'After much further research by a number of theoretical physicists, I am more convinced than before that macroscopic wormholes, large enough for humans to travel through, do not occur naturally in our universe. But I'm not sure. It might be that very advanced civilizations can make wormholes artificially, but if so, I think it likely that the laws of physics prevent such civilizations from collecting enough exotic material inside a wormhole to hold it open. This means that nothing can ever travel through a wormhole from one side to the other. But I'm not sure.'

Despite the uncertainty over the feasibility of wormholes, microscopic versions, small enough to pop up and disappear in a short space of time, do turn up in several advanced theories at the front line of physics, as a means of solving theoretical questions such as the information paradox.[4] Wormholes are one common answer to questions about what happens inside a black hole. Does the journey into the darkness lead to a portal or passage that continues out the other side? Could this idea, familiar from science fiction, also be realized in nature? When Event Horizon

Telescope scientists released the first image of Sagittarius A* in 2022, they investigated whether this object might be a wormhole.[5] By calculating what the shadow of such a wormhole would look like, they concluded that it would have been smaller than the object they'd observed. The existence of wormholes is therefore not only a theoretical question, but also an empirical one.

A MANTIS SHRIMP'S VISION

A few months before the image of Sagittarius A* was released, I took a cable car up through the French Alps, hiked through the snow along a precipice overlooking a several-hundred-metre drop, and got a lift on a snowcat. At last I reached the Northern Extended Millimeter Array observatory and its twelve telescopes, all situated on the Plateau de Bure. In the observatory's canteen I listened to the astronomer Michael Bremer. He had just completed some new observations of Sagittarius A* as part of the Event Horizon Telescope team, and as we drank coffee he offered a new perspective on black holes that surprised me.

'The mantis shrimp can see twelve primary colours,' Bremer said. 'Humans, on the other hand, can only see three. Our senses do not perceive anywhere near what is possible in nature. But thanks to our intelligence and technology, we can expand and enhance our senses. We can observe things that we could not see or experience before.'

Like black holes. Through a window I catch sight of one of the telescopes, pointed up at the blue sky. Beyond it, the spring sunshine beats down over the snowy mountain ranges that stretch all the way to the horizon. The telescope is a part of humanity's extended vision. With the aid of a frequency divider, fibre-optic cables, super-cooled radio receivers and telescope dishes that can be angled by the micrometre, it's possible for Bremer and the other astronomers at the Event Horizon Telescope to depict black holes.

The mantis shrimp might be able to see more basic colours than a human, but they can't build radio telescopes and take photos of black holes. But in spite of humans' technological ingenuity, which takes us beyond our sensory limitations, there are boundaries to how much we can discover. In the process of writing this book I've met countless people who've contended with the theoretical analysis or empirical observation of black holes, and to each of them I've put the same question: if you could travel to a black hole and explore it, what would you want to find out? Most of them have answered in the same way as Shep Doeleman from the Event Horizon Telescope team: 'I would like to go into a black hole to see what happens to space and time inside.'

I agree. I've often tried to imagine what it would be like to fall into a black hole and survive the trip long enough to see what its darkness holds. It's still unclear what happens to the matter that enters a black hole. There's no simple answer to the question of what the end point of matter is when gravity is allowed to act unchecked. Unfortunately, Jean-Pierre Luminet was right when he said, 'The only thing we can send into a black hole is our equations.'

Something a number of scientists have examined using these equations is whether black holes can harbour whole new universes.[6]

A NEW UNIVERSE

In the late eighties the Russian physicists Valeri Frolov, Moisey Markov and Viatcheslav Mukhanov published an article entitled

'Through a Black Hole Into a New Universe?'[7] The three physicists are highly respected, and Mukhanov's calculations of the very early evolution of the structures of the universe via quantum fluctuations are a cornerstone of modern cosmological science.

In their article, they suggested that when matter collapses and creates a black hole, a new universe can be created inside it. The three physicists based their theory on an assumption: as matter is squeezed together inside the black hole, the curvature of spacetime increases more and more. Eventually it can be infinitely large. But according to quantum mechanical theory, this ought to bring about new physical effects, as space and time change drastically at the tiny distances of the quantum mechanical world. When Frolov, Markov and Mukhanov analysed what happens when there is an upper limit to how much spacetime can change, they discovered that a natural consequence is the creation of a whole new world, similar to a new universe, inside the black hole. They wrote that it should be possible "'to travel" from our universe into a new one'.

A number of physicists have investigated the idea of a new space that resembles a new universe being created inside a black hole. Viewed from the outside, the black hole's size would be unchanged. This situation demonstrates what an apparently contradictory situation spacetime curvature can lead to: an object has a constant size from the outside, while on the inside, a new universe is taking shape.

In purely mathematical terms, this is possible. As previously mentioned, Kerr's formula for rotating black holes contains not only a new universe in its interior, but *an infinite number of possible new universes*. These appear to be connected via event horizons inside the black hole. It is as though Roy Kerr discovered the formula for a world that holds many worlds. But the existence of a mathematical formula that describes a phenomenon doesn't necessarily imply that the phenomenon exists in nature. Kerr's formula described a rotating black hole that existed for ever. It has always existed, and will always continue to exist. The black holes found in nature, however, have been created in the explosions of stars or the collapse of gas clouds. That's why physicists must use new models to study the theoretical possibilities of what can happen at the end of such a collapse. The idea posited by Frolov, Markov and Mukhanov is one such model of how a new universe can be created during the formation of a black hole. Some have even argued that our own universe could be born out of a black hole.

One of them is the US physicist Lee Smolin.[8] He has taken the idea of new worlds inside black holes to its extreme. In the early nineties he suggested that every time a black hole is formed, it gives rise to a new universe. There, the mathematical values that determine the properties of nature can vary slightly. Perhaps the gravitational force is somewhat higher, or electrical charge a little lower. Perhaps light travels slightly faster than in our universe. In total there are approximately twenty physical constants of nature whose values might shift when a new black hole, and a new universe, come into being. In some of them, planets, stars and galaxies form, while in others, nature works so differently that completely new physical processes arise. The existence of black holes therefore leads to a multiplicity of possible worlds, in which all the worlds look somewhat different to the one from which they were created.

In our observable universe, every second sees the explosion of roughly a hundred stars that could lead to the formation of a black hole. If Smolin's theory proves to be true, this would mean the creation of around a hundred new universes every second. Because of the variation in the physical constants, the probability of new black holes appearing in these new worlds varies. This leads to a kind of evolution of these new worlds: in the ones containing the most black holes, even more new worlds will in turn be created, which again could potentially lead to the creation of even more black holes.

Inspired by the process of evolution on Earth, Smolin called this process *cosmological natural selection*. Evolutionary biology tells us that small changes in an organism's genetic material can lead to it being able to maximize its chances of surviving and passing on its genetic material. Similarly, worlds that maximize the number of possible black holes will form an increasing proportion of all possible worlds.

The aim of Smolin's argument was to explain why the fundamental constants of nature, which govern phenomena such as the gravitational force or the charge of electrons, have the values they do. *If* new universes are formed inside black holes, it's most likely that we live in a universe that's optimized for creating as many black holes as possible. According to this radical suggestion, the number of black holes in our universe would be relative to the values of the physical constants. Smolin had an additional aim, however. Many physicists have been fascinated by how finely tuned these constants seem to be to the existence of life. If they changed by a fraction, natural processes would be so different that life, at least as we know it, could not exist. This has given rise to a number of philosophical discussions – which not infrequently turn in a religious direction – around why it is possible for life to occur in the universe. Smolin wanted to explore an alternative explanation that instead considered the fact that these constants are conducive to the formation of black holes. In a way, this idea brings the relationship between black holes and the origin of life full circle: the existence of the former proves advantageous to the latter.

ARE WE LIVING IN A BLACK HOLE?

It's certainly an intriguing thought, that the black holes in our galactic neighbourhood, closer to us than some of the stars in the night sky, could be portals to other universes that function completely differently to ours. But it's impossible to tell whether Smolin's theory holds, for the simple reason that we cannot know what happens within a black hole's event horizon.[9] Still, there are several hypotheses like Smolin's that explore the possibility of new worlds occurring inside black holes, and the fact that these hypotheses exist shows how these objects have become theoretical laboratories for the most advanced – and speculative – ideas about the properties of nature and the universe. The logical extension of these models leads to a dizzying question: could the entire universe be a black hole?

THE COSMIC EVENT HORIZON

'New Nasa data hints we could be living inside a black hole,' exclaimed the headline of the *Independent* newspaper on 19 March 2025.[10] A computer scientist had analysed data from the James Webb Space Telescope, and, based on a study of how galaxies rotate, made the bold statement that the universe could, indeed, be created from within a larger black hole.[11] The idea that our universe is formed within a black hole is supported further by similar claims in the scientific literature. Could this really be true?

There is an important argument that often crops up in discussions as to whether our universe is a gigantic black hole.[12] The argument builds on the most fundamental characteristic of a black hole, which is a consequence of the formula Schwarzschild wrote down to the accompaniment of the thudding artillery guns on Hartmannswillerkopf: that the mass of a black hole determines the size of its event horizon. In order to determine whether the universe might be a black hole, we must therefore find out how much mass there is in the universe and how big it is.

It is tricky to determine the size of the universe, because of one of the greatest astronomical discoveries of the twentieth century: that the universe is expanding. In fact, this finding was so revolutionary, it would be fair to call it one of the greatest scientific discoveries in history.

The expansion of the universe means that all distances are growing over time. The universe was once smaller, it has grown and it will continue to grow in the future. The universe is therefore not in a constant state; the whole cosmos is always changing. There is no eternal state of equilibrium – neither for the continental plates, the Sun, the galaxies nor the universe in its entirety.

This revelation came about in the context of an intense early-twentieth-century debate about the size of the universe. Several astronomers had begun to wonder whether anything lay beyond the Milky Way and whether there might be other Milky Ways scattered across the cosmos. Today we take it for granted that the universe is teeming with galaxies, but it's fascinating to think that at the time my grandparents were born, many astronomers believed that the Milky Way was all there was. Our understanding has expanded dramatically in the space of just a couple of generations.

In 1924 the astronomer Edwin Hubble shook up the debate about the size of the universe.[13] Astronomers had been wondering whether the great star system Andromeda was a galaxy or just a system within the Milky Way. In order to find out, Hubble realized he needed to know the distance to these stars. He wasn't the first to attempt this; others, including the Swedish astronomer Knut Lundmark, had previously argued that it must be a galaxy in its own right. But Hubble had access to data from a newly built telescope on Mount Wilson in California. He also

ARE WE LIVING IN A BLACK HOLE?

used a new way of measuring cosmic distances introduced by the astronomer Henrietta Swan Leavitt.

In the early years of the twentieth century Leavitt had an unpaid position at Harvard University, where she was responsible for astronomical calculations as one of Harvard's team of human 'computers'. Although she was not initially employed as a researcher, her familiarity with the university's star catalogues meant that she later came up with a new method of calculating the distance to a particular kind of star known as a Cepheid variable. Leavitt realized that the strength of the light from these stars changed with a certain regularity. The length of time these regular changes took depended on the total quantity of light they produced. Therefore, it was possible to determine how far away they were by measuring how much of their light was visible from Earth and how long it took that light to increase and decrease in strength. Hubble employed Leavitt's discovery in his observations and was able to confirm that Andromeda was more than nine hundred thousand light years away – a distance much larger than the size of the Milky Way (modern measurements put the distance at approximately 2.5 million light years). Hubble's results gave the first convincing proof that the speculations about Andromeda were correct. The Milky Way was not the entire universe – there were other galaxies that occupied its vastness.

The decisive confirmation that Andromeda is a galaxy like the Milky Way was accompanied by the realization that many galaxies (although not Andromeda) are moving away from us. This led to the discovery that our universe is expanding, a breakthrough made possible by research on two fronts. The first was theoretical: many scientists had explored what Einstein's general theory of relativity meant for the properties of the universe. Einstein himself had initially insisted that the universe must be static, and had made theoretical manoeuvres to dodge one of the most important predictions of his theory (which he later, according to the physicist George Gamow, came to call his 'biggest blunder').[14] But several scientists demonstrated that his theory really did predict that the universe was expanding. The other front was empirical. Many scientists were analysing the new data facilitated by improvements in the telescopes and instruments used to study the structure of cosmic light. It was this encounter between the mathematical analysis of the general theory of relativity and improved observations of the properties of the light reaching us from distant galaxies that led to the discovery that the universe is expanding. The discovery is often attributed to Edwin Hubble alone, but there were also contributions from Knut Lundmark, the US astronomer Vesto Slipher, the Belgian priest and scientist Georges Lemaître and many other astronomers and physicists.

So, what did this discovery mean? When astronomers analysed the light emitted by galaxies, they saw that its wavelength had got longer. This phenomenon is called redshift, and we encountered it in Chapter 5. The further away a galaxy was, the more redshifted its light appeared to be. This was easy to explain by taking the expansion of the universe into account: when light travels through expanding space, its wavelength gets extended by the expansion. However, the expansion of the universe does not happen in the same way as an object growing and gradually taking up more space, like a balloon being inflated, or an explosion spreading from the point of detonation. Instead, it is space itself that gets bigger. Nor is the universe expanding from a central point; instead, everything is simultaneously moving away from everything else. On Earth we don't notice this expansion, and nor can it be observed within the solar system, the Milky Way or even in relation to our closest galaxies. The Andromeda galaxy is bound to the Milky Way by gravity, and the two are travelling towards one another rather than away. It's only at a cosmic scale that the expansion of space becomes noticeable.

The realization that the universe is expanding made humanity reconsider once again its place in the cosmos. It's a shift that began with Copernicus's suggestion that the Earth is not at the centre of the solar and planetary orbits. Then it turned out that the solar system is not at the centre of the Milky Way, and that the Milky Way is not even at the centre of the universe, but merely one of more than two thousand billion galaxies. The final step in this displacement of humanity away from the centre was the complete dissolution of the idea that the universe even has a centre.

We do not know how big the universe is right now. It could be infinitely large, or its size could be finite. It takes time for light to travel from one point to another, and since the universe is around 13.8 billion years old, there's a limit to how far we can see. It's for this reason that we can't find out the universe's full size.

The oldest light we can see comes from the cosmic background radiation, which started to stream freely through the universe about 380,000 years after the Big Bang. Because the universe is expanding, the starting point of a ray of cosmic background light reaching us today would now be about 46.5 billion light years away from us.[15] Even though the light ray has been travelling for about 13.8 billion years, since space itself has stretched during this time, the region where the light originally came from is now much further away. This distance defines our observable universe, which has a diameter of ninety-three billion light years. This is probably only a small part of the universe's total size.

If we use this definition of the size of the observable universe, we can begin to investigate whether our universe could be a black hole. To find this out, we need to know how much mass this volume contains. When astronomers look out into the universe, they see that it is mostly empty. There are, of course, areas with a high concentration of matter, such as galaxies, stars and planets. But the distances are so vast that the universe is, on average, a pretty desolate place. Indeed, its expansion is making it ever-sparser and emptier.

Finding out exactly *what kind* of matter and energy the universe contains is one of the most important tasks of modern cosmology. The contents determine both how the universe has expanded in the past, and how this expansion will continue in the future. A range of observations of astronomical phenomena have contributed to the standard model of cosmology, which states that the universe consists of approximately 5 per cent ordinary matter, 27 per cent dark matter and 68 per cent dark energy.[16]

Exactly what constitutes the latter two forms of matter and energy is, at present, unknown. We learned a bit about dark matter in Chapter 9, and we're going to find out about the implications of dark energy shortly. But first, we need to answer the question of the universe's mass. When astronomers add up all the contents of the observable universe, they calculate the amount of mass at around 10^{54} kilograms (1 followed by 54 zeros).[17]

With knowledge of the mass and size of the universe, it is possible to perform the same calculation astronomers do when they want to find out whether a celestial body is a black hole. The Schwarzschild metric gives a value for the volume a body with a given mass would have if it *were* a black hole. If the body's size is roughly the same size, it's very likely to be a black hole. For the Earth, we saw that its mass would have to be pressed into a sphere with a diameter of a little less than two centimetres. For the Sun the equivalent diameter would be six kilometres, while for the Milky Way, the value would be about 0.4 light years. This is as far as the distance from the Earth to

ARE WE LIVING IN A BLACK HOLE?

parts of the Oort cloud, the spherical shell of ice particles that surrounds the solar system.

If we use Schwarzschild's formula for the entire mass of the universe, it states that a black hole of that mass would have a diameter of about three hundred billion light years.[18] That's more than three times the size of our observable universe! In this sense, then, the answer to the question of whether we live inside a black hole is yes.

The argument is supported by a surprising fact: the universe has an event horizon. During the twentieth century, about half a century after Hubble and other astronomers had established that the universe is expanding, astronomers confirmed that the pace of that expansion is increasing over time. This result was based, among other things, on careful analyses of supernovas that exploded several billion years ago.

Dark energy is the energy form driving this acceleration. The adjective 'dark' refers to the fact that this energy form cannot be seen directly, and its origin is unknown.[19] It's possible that dark energy is connected to the energy content of the vacuum itself.

Because of dark energy, galaxies that are a certain distance away from us will travel so fast away from us that their light will never be able to reach us, even if we were to wait patiently for a very long time.[20] If that sounds familiar, it's because it harks back to the situation with black holes, where the light from inside the black hole will never reach us regardless of how long we wait.

This means that the universe too has an event horizon, beyond which we can never know what happens. Right now, several galaxies are passing this horizon. Far off in the darkness of space, the galaxies are travelling away from us so rapidly that their light is fading from view. We will never see them. In the future, in dozens of billions of years, dark energy will have driven all the universe's galaxies beyond the cosmic event horizon. Only the few that are gravitationally linked to the Milky Way in the local group will be visible. The possibility of gaining knowledge about a huge part of the universe will be lost.

That the universe has an event horizon and that a black hole with the same mass as the universe would be bigger than our observable universe are the main reasons that several scientists take seriously the idea that we may actually be living inside a black hole. But there are three phenomena that contradict this idea.

1. Inside a black hole, space changes in a peculiar way over time. In one direction it is stretched out, in the other it is squeezed together. Think back to our astronaut falling into a black hole and getting drawn out in one direction and squeezed in the other. In our universe, however, space is expanding at an equal rate in all directions. The properties of space in the universe therefore bear no resemblance to those same properties in a black hole.

2. Inside a non-rotating black hole there is a singularity – the point at which spacetime curvature is infinitely large – in the future. In the Big Bang model of the universe, there also appears a singularity. When we extrapolate the expansion backwards in time, all forms of matter and energy will eventually be squeezed together so tightly that the spacetime curvature becomes infinitely large. This is the hallmark of a singularity, but unlike black holes, in the Big Bang model this singularity lies in the past.

3. Light and matter outside a black hole can fall in through the event horizon and reach its interior. In the universe it's the opposite: light and particles beyond the cosmic event horizon will never reach us. The universe's event horizon is therefore

a kind of inside-out version of a black hole's event horizon. What's more, the cosmic event horizon will depend on the observer. A being in a far-away galaxy would have a different event horizon to us. The cosmic event horizon is relative, while the event horizon of a black hole is a fixed surface in spacetime.

These three phenomena – the expansion of space, the singularity and the event horizon – differ so significantly between a black hole and the universe that any similarities between them break down.[21] What's more, the universe contains a certain quantity of mass and energy, while the mathematics of a black hole describe a vacuum.[22]

What about the media claim, then, in March 2025 that we could be living inside a black hole? As so often happens, such far-reaching science claims spread like wildfire in both traditional and social media. But the science case is more complex than the headlines indicate. For the particular study in question, there was no clear link between what was claimed to have been observed by the JWST (an asymmetry in the direction of rotation in a small sample of galaxies), and how this is evidence that our universe is embedded in a larger black hole. Other more mundane explanations of the claimed asymmetry, such as data selection effects or astrophysical properties of the galactic rotations, were far more likely.[23]

Even if the universe is probably not a black hole, there are important lessons to be learned from the fact that they both contain event horizons. In a way, they are one another's opposites. Black holes are created when matter is drawn inwards and implodes, while the universe's expansion causes matter to become sparser. In spite of this difference, event horizons arise both when matter collapses and as the universe expands.

It's as though the universe will not allow us to know everything that is happening inside it. The dream of science is to understand and explain the entire cosmos. But because of the event horizons of both the universe and a black hole, an absolute limit is placed on our access to knowledge. It's not a question of our instruments being insufficient; nature itself sets these limits. One of the most important scientific discoveries ever made about the universe is therefore not about a specific object, but about *a limit to knowledge*. In the case of black holes, how these limits change over time depends on the exact characteristics of their Hawking radiation, and in the case of the universe, on the long-term evolution of dark energy. Trying to understand these two phenomena is one of the most challenging, but important, problems of modern physics. But even if these problems are not yet completely solved, we can still use the best current cosmological knowledge to conjecture what will happen with the universe and black holes in the distant future.

THE END OF THE UNIVERSE

In the future, the universe will be full of darkness. All the stars will go out and no new ones will be able to form, because the supply of dust clouds will be gone. Like our Sun, most stars will have become white dwarfs, radiating away their inner heat over billions of years and growing dark and cold. The same will happen with the universe's neutron stars.

Extrapolating further, it is likely that all the billions of black dwarfs, neutron stars, planets and millions of black holes in every galaxy will either move towards the centre of their galaxies and the supermassive black holes there, or be flung out of the galaxies through gravitational interactions. Across inconceivably

ARE WE LIVING IN A BLACK HOLE?

long timescales – around 10^{24} years (1 followed by 24 zeros) from now – the supermassive black holes will therefore engulf many of the objects that make up the galaxy.[24] Some of the galaxies that have not moved away from one another in the expansion of the universe may even collide with each other, whereupon the supermassive black holes at their centres may merge.

At the same time as the supermassive black holes are growing ever larger, the universe will expand ever faster. The light that remains in space will become fainter and fainter as it is stretched thinly by the expanding cosmos. Eventually, the galaxies will travel beyond the limits of what we can see. The universe will be vast, dark and empty, full of supermassive black holes. It will be a place where it is impossible to get information about the make-up of the universe, not only because of the darkness, but also because of the event horizons of both the cosmos and the black holes.

But even these black holes will ultimately disappear, slowly shrinking because of their Hawking radiation. The timescales are so long they are difficult to imagine. Perhaps the characteristics of the universe will have altered fundamentally – according to some models, the whole universe might have been torn apart long before the last black holes have evaporated. But if the universe is still in existence 10^{68} times (1 followed by 68 zeros) longer than it has hitherto existed, the black holes created from the death of stars will have radiated away. At even longer timescales, even the supermassive black holes will shrink and vanish. They were some of the first objects created in the universe, and they will be some of the last objects left in that distant future, 10^{98} years away. When they're gone, the universe will be filled only with the faint, ever-waning glow of the black holes' remaining Hawking radiation. No life will exist, no ideas will be thought up, no matter will remain. By all accounts, the last stage of the universe will be a dark and constantly growing void.

THE MOUNTAIN OF LIFE

More than two hundred years ago, the British rector John Michell wondered whether there might be stars whose gravity was so strong no light could leave them. More than a hundred years ago, the German astronomer Karl Schwarzschild discovered the mathematical formula for black holes. Since then, physicists and astronomers have worked hard to try to understand the consequences of the formula contained in the letter that Schwarzschild sent to Albert Einstein. A little over a hundred years after it was written, it has taken us to the very boundaries of knowledge and the universe.

I emerge from a tunnel on Hartmannswillerkopf and am dazzled by the sunlight. I squint, looking at the barbed wire coiled above the mouth of the tunnel. At the start of the First World War, the mountain was covered with thick forest. By the end, the trees had been blown to pieces, and ropeways, trenches, walls, roads and tunnels criss-crossed the mountain. 'I have experienced hours that felt like centuries,' a soldier who fought in the trenches said, 'where death brushes against you every second.'[25]

It is telling that the formula for black holes was discovered in close proximity to the Mountain of Death. Several of the mountains and space bases involved in black hole research are situated in locations marred by colonial conflict, very often over who has the right to decide what happens on a mountain. Moreover, death is one of the most common associations black holes raise. Perhaps the association is so prevalent because of that ancient idea of death as a kind of subterranean imprisonment that appears in many mythologies. The souls that end up in the kingdom of death are forever caught in the darkness of the underworld. Similarly, no one who passes the event horizon can ever return. But someone who finds themselves inside a black

ARE WE LIVING IN A BLACK HOLE?

hole can continue to exist for a time, experiencing what happens in the darkness. When I interviewed the Canadian astrophysicist Avery Broderick from the Event Horizon Telescope team, he said, 'I've heard that story before. I heard it in Sunday school. You die and you continue to exist, but you can't tell anyone. It's the same kind of story.'

The symbolic power of black holes is as great as their gravitational force. But as our knowledge of them grows, we must go beyond our spontaneous associations and replace them with new ones. In the Hawaiian creation story, darkness has an important role. It is to be found everywhere, and is, by extension, the origin of all life. Equally, astronomers have learned that black holes have a creative role in the universe. They influence the evolution of galaxies, are behind one of the brightest lights in the universe, and, as we've seen, can even affect the possibility of life occurring in different locations in the cosmos. When the astronomer Jessica Dempsey began collaborating with the activists seeking to defend the mountain Maunakea from further exploitation, she realized that black holes even had the potential to change the relationship between astronomers and the inhabitants of the Hawaiian islands.

When I visit the Mountain of Death I see a parallel to the new images of black holes. As I wander around the trenches, I hear the twittering of birds and the chirping of crickets among the ash, lime and fir trees. Insects fly about between the bushes and flowers. Woodpeckers eat the insects, and bats have found a safe haven in the abandoned military tunnels. Hartmannswillerkopf was designated a Natura 2000 site by the European Union because of its unique ecology. Rare species, such as the flower rosy saxifrage and the Jersey tiger moth, flourish in the silicon-rich environs. New life abounds. The Mountain of Death has become the Mountain of Life. In a similar way, it's time for us to expand the range of things we associate with black holes, from death, darkness and desolation to life, light and creation.

If I could choose a word to describe black holes, it would be *duality*. They are the universe's darkest objects, but they are behind its brightest light. They are distant, and we will probably never even get close to one, and yet they are used to calibrate our global navigation systems and help us understand our own planet. They are associated with death, but astronomers have realized they could be a prerequisite for life. They are a kind of absolute end point for space and time and yet, in several models, they are also passages to new universes. For a long time, scientists believed they were completely black, but Stephen Hawking showed that they can radiate. In one sense, they are the simplest objects in the universe, described through only their mass, rotation and, in some cases, electrical charge, but in another sense they are the most complex objects, with the highest entropy. They can contain the most mass inside a given volume, but consist, simultaneously, of nothing but space and time.[26] Black holes are, quite simply, a kind of unity of opposites. I cannot think of any other object – if they can even truly be called objects! – of which the same could be said.

There is much we do not know about black holes. They are almost as mystical now as they were when they were discovered. We have come to realize that to understand them requires answers to some of the most fundamental questions we can pose: whether information can be lost in nature, how galaxies come into being and stars die, how our fundamental theories of nature can be united, and whether there are portals to different locations in space or other universes. All the same, the foremost question concerning them remains to be answered, and perhaps always will do: what happens at the very centre of a black hole?

ARE WE LIVING IN A BLACK HOLE?

I contemplate this question as I'm driving back along the winding roads of the Vosges mountains. The Sun is low in the sky, shining its golden light on vineyards and villages. I still struggle to imagine how a mathematical formula that was written down in close proximity to this place could have been the starting shot for a scientific journey we have still only seen the beginning of.

Over the coming years, more experiments will give us new insights into black holes. Already within the first few months of 2025, as I'm making the final edits of this book, a series of new results about the dynamic activity of black holes has been reported: astronomers have discovered new infrared flares breaking out around Sagittarius A* as well as a three-day-long outbreak of gamma-rays around Pōwehi; they have seen a black hole launching jets in real time and carefully studied how such jets accelerate in sixteen active black hole systems.[27] Hopefully, there will be new telescopes this decade, gravitational wave detectors in the next, and large radio telescopes in space and on the other side of the Moon by the middle of this century, all of which will help us answer more questions, such as exactly how many black holes there are, what properties they have, where they come from – and what role they play not only in the universe, but for humanity's place on Earth.

ACKNOWLEDGMENTS

It took me five years to write *Facing Infinity*. During this time I've had the pleasure of being helped by and getting to know many people. First of all I would like to thank my Swedish publisher Kerstin Almegård and my editor Erika Palmqvist at Albert Bonniers Förlag. It's been fantastic watching them help me transform a sprawling mass of text into a finished book. Thanks also to my agent Steve White at Nordin Agency for his constant enthusiasm. For the English edition, I'm grateful to my translator Nichola Smalley for her careful work, my copy-editor Tamsin Shelton for her meticulous editing and my publishers Ed Faulkner and Matthew Lore for putting their trust in this book.

Interviews form a central part of this book. Huge thanks to all those I have interviewed, be that in person, by video call or over email: Nils Andersson, Gianfranco Bertone, Geoff Bower, Michael Bremer, Avery Broderick, Chiara Ceccobello, John Conway, Jessica Dempsey, Shep Doeleman, Sheila Dwyer, Heino Falcke, Reinhard Genzel, Andrea Ghez, Ariel Goobar, Rüdiger Haas, Carl-Johan Haster, Sara Issaoun, Sam Keliiheleua, Roy Kerr, Larry Kimura, Michael Landry, Jean-Pierre Luminet, Priyamvada Natarajan, Mark Pearce, Juan Peñalvez, Alain Riazuelo, Klas Rönnbäck, Aurora Simionescu, Douglas Simons, Kip Thorne, Salvatore Vitale, Marta Volonteri and Rainer Weiss. Some of them provided vital background information, others are featured in the book. Many of those I've described have also read and commented on the text. I was especially glad that the Nobel Prize laureate Rainer Weiss made time to do this.

Many people have fact-checked different parts of the manuscript. Thanks to them the book has improved – but any remaining mistakes are my own. In particular I want to thank Carl-Johan Eberstein, Rebecca Forsberg and Nikki Arendse for their extensive feedback, as well as Angela Adamo, Jim Bayman, Ingemar Bengtsson, Robert Cumming, Daìnis Dravins, Rikard Enberg, Ariel Goobar, Ulf Gran, Carl-Johan Haster, Matthew Hayes, Sören Holst, Joel Johansson, Johan Kärnfelt, Tim Linden, Joakim Mansfeld, Stuart McAlpine, Jens Melinder, Edvard Mörtsell, Christian Ohm, Stephan Rosswog, Felix Ryde, Martin Sahlén, Emilie Skulberg, Sverker Sörlin and Erik Zackrisson.

I express my acknowledgment and gratitude to the original stewards of the dispossessed locations I've visited while working on this book: in Sápmi, the traditional land of the Sámi people; on Hawai'i, as a guest of the Kānaka Maoli, and their 'āina; at Hanford, on the traditional land of the Walawalałáma, Imatalamłáma, Yakama, Wanapum, Weyíiletpuu, Pelúutspuu and Nimíipuu peoples; and at Cambridge, traditional home of the Massachusett people.

A huge thank you to the staff at NOEMA, IRAM 30-meter, LIGO Hanford, Submillimeter Array, Onsala Space Observatory, South Pole Telescope and Esrange, for letting me visit these unique sites. I would also like to thank the staff at the Göttingen State and University Library, the 'Imiloa Astronomy Center, and Geoff Bower, who guided me at the observatories on Maunakea.

Thanks to the members of the History, Philosophy and Culture Working Group at the Next Generation Event Horizon Telescope, in particular Peter Galison, Niels Martens, Juliusz Doboszewski, Jamee Elder, Ann C. Thresher and Daniel Palumbo. Our discussions have enriched this book. Emilie Skulberg, who is also part of the group, deserves extra gratitude for all her support and conversation.

ACKNOWLEDGMENTS

I've benefitted hugely from the opportunity to debate ideas with colleagues. Thank you to my former colleagues at the Oskar Klein Centre, the European Southern Observatory's communication division in Garching, especially Juan Carlos Muñoz-Mateos, and the Vossius Center for the History of Humanities and Sciences in Amsterdam, especially Jeroen van Dongen and Rens Bod. Thanks to Sera Markoff for many rewarding discussions and for inviting me to take part in her research group at Amsterdam University.

Thanks to Ignatius Adriaan Raap, who shared his materials relating to Karl Schwarzschild, and with whom I've had the pleasure of starting a research collaboration. Thanks to Sören Holst for all the valuable conversations and popular science collaborations.

In addition to these people I would like to thank those who have contributed in large and small ways to my work on this book: Thors Hans Hansson, Maria Küchen, Robert Cumming, Jonas Ask and all the teachers and students on the narrative non-fiction course at Jakobsbergs folkhögskola, Alexandra Aldea, Inger Lundin, Fanny Nilsson, Johan Härnsten, Ida Hansen, Linnea Garli, Hanna Asp, Merl Fluin, Katrin Ros, Emma Lundenmark, León Sosapanta, Kālewa Correa, Ingalill Thorsell and the participants in Drakamöllan Nordic Forum for Culture and Science.

Articles that form the basis for parts of this book were previously published in *LiveScience, ETC, Glänta, Arbetaren, Populär Astronomi* and *Forskning & Framsteg*. Thanks to all the editors who have worked on my texts.

I am grateful for the support of the Helge Ax:son Johnson Foundation and Drakamöllan Residency Fellowship.

Thanks to my family who've always been there for me: Katinka, Jesper, Oskar, Elias and Kicki.

And a final thank you goes out to Alma who, as an eight-year-old, sent me five questions about black holes when I was just starting to think I should give up writing. Those well-formulated questions set me back on the right path – and here we are, with a finished book.

NOTES

PROLOGUE

1 Pascal (1670).
2 The journey into a black hole is based on the mathematical formula for a non-rotating black hole, and the system of coordinates introduced in the twenties by the Frenchman Paul Painlevé and the Swede Allvar Gullstrand. I was inspired by the scientific simulations conducted by Alain Riazuelo, a researcher at the Institut d'Astrophysique de Paris. See also Matzner, Rothman and Unruh (1985) and Taylor and Wheeler (2000).

CHAPTER 1

1 Sicilia et al. (2022).
2 My description of Michell's life is based on several sources, primarily McCormmach (2012), but also Hardin (1966), McCormmach (1968), Hutton (2006), Hughes and Cartwright (2007), Montgomery et al. (2009) and Geikie (1918). See also Carter (1987), Israel (1989), Eisenstaedt (1991) and Bartusiak (2016).
3 This phrase was uttered by John Playfair when presenting Hutton's research. See Stone (2012).
4 Laplace (1796), translated from the French by Henry H. Harte.
5 Michell (1784).
6 Hankins (1985).
7 More specifically, Newton showed that the gravitational force between two bodies decreases with an inverse square of the distance.
8 Michell (1784). It is often said that Michell coined the expression 'dark stars' for these dark celestial bodies, but as far as I know, the expression does not appear in his writings. The term has subsequently become the established one by which to refer to his idea, and I use it in this sense.
9 Letter to William Watson, 1781, cited in McCormmach (2012). Another natural philosopher, who lived in York, complained that 'there is not a soul here to converse with'.
10 McCormmach (2012).
11 Letter to Henry Cavendish, 1783, cited in McCormmach (2012).
12 Eisenstaedt (1991), see also Carter (1987).
13 Laplace (1796). He presented a mathematical proof for dark celestial bodies in Laplace (1799). Instead of basing it on the average density of the Sun, as Michell had done, Laplace showed that a celestial body would be completely dark if it was two hundred and fifty times the size of the Sun and had the same density as Earth.
14 McCormmach (2012).
15 Von Soldner (1800). He also wrote, 'I am, for the most part, in agreement with Mr Laplace that there may be heavenly bodies that do not emit any light, but these could, to my mind, equally well be small as large.' This is, as far as I am aware, the first time a scientist speculated that there could be small, and not just large, heavenly bodies that are dark because of their strong gravity.

16 McCormmach (2012).

17 Von Soldner (1800); see also Israel (1989).

18 Michell suggested that variations in the speed of light from two nearby stars could be measured using a prism. Cavendish conducted an experiment to search for variations of this nature, but did not find any. Alongside investigations into the wave properties of light, these experiments increased doubts about the existence of dark celestial bodies.

19 https://www.nobelprize.org/prizes/physics/2020/

CHAPTER 2

1 Lynden-Bell and Rees (1971).

2 See Sanders (2014) for a detailed historical description of how the astronomers learned more about the centre of the Milky Way.

3 See Becklin and Neugebauer (1968), and Balick and Brown (1974).

4 Oort and Rougoor (1960).

5 The radio astronomer and Nobel laureate Charles H. Townes was instrumental in the observations conducted to establish whether there was a supermassive black hole at the centre of the Milky Way. By studying the speed of the gas cloud, he and his colleagues demonstrated that Sagittarius A* was as heavy as several million Suns. See Wollman et al. (1976).

6 https://www.scientificamerican.com/article/how-andrea-ghez-won-the-nobel-for-an-experiment-nobody-thought-would-work/

7 Genzel and his colleagues determined, among other things, the distance to the centre of the Milky Way; see Eisenhauer et al. (2003).

8 The Very Large Telescope was constructed and is operated by the European Southern Observatory (ESO). Genzel and his colleagues initially used the ESO's New Technology Telescope at the La Silla Observatory in Chile.

9 https://www.spiegel.de/international/zeitgeist/nobel-prize-in-physics-winner-it-s-unbelievable-all-that-s-going-on-at-the-moment-in-astronomy-a-6b4867e5-ead6-43f3-910b-bf69e9485ff2

10 Galileo (1610).

11 Ghez et al. (1998).

12 Jia et al. (2019).

13 Ghez et al. (2003) and Gravity collaboration (2018).

14 My portrayal of Brahe and Kepler's encounter follows Gilder and Gilder (2005).

15 Letter from Andreas Foss to Brahe's assistant Longomontanus; see Gilder and Gilder (2005).

16 Berggren (2009) and https://phys.org/news/2012-11-mercury-poisoning-tycho-brahe-death.html

17 Data concerning the planets has been retrieved from https://solarsystem.nasa.gov/planet-compare/

18 Kepler noted, for example, that Jupiter's moons had an orbital period and a distance to Jupiter from which it was possible to derive a common value by repeating similar mathematical operations. This value depended in turn, according to Newton, on the mass of Jupiter.

19 A detailed presentation of Ghez and Genzel's scientific contributions can be read in the Nobel Prize Committee's advanced report; see https://www.nobelprize.org/prizes/physics/2020/advanced-information/

NOTES

CHAPTER 3

1 The folders are in the Göttingen State and University Library under the shelfmark 'Cod. Ms. K. Schwarzschild'.

2 The passages describing Schwarzschild's life are based on several sources; see Schemmel (2004), Raap (2017) and Raap (2024). I would like to thank Adriaan Raap for our collaboration on this book and on an article that recounts Schwarzschild's activities during the First World War. Raap has transcribed several of the close-to-indecipherable letters that can be found in the archive at the University of Göttingen, and has been very generous in granting me access to this material.

3 Schwarzschild used the German spelling Hartmannsweilerkopf, but I use the French spelling.

4 Schwarzschild (1920).

5 Letter to Else Schwarzschild, 19 December 1915.

6 https://einsteinpapers.press.princeton.edu/vol8a-doc/296

7 The descriptions of Einstein's life in this and the next chapter are primarily based on Levenson (2004), Pais (2005), Hoffmann (2006), Isaacson (2008) and Neffe (2021). Galison (2003) is a fascinating portrayal of how technology and politics influenced Einstein's thinking. See also Miller (1998) and Miller (2002).

8 https://www.thenation.com/article/archive/myth-new-anti-semitism/

9 Neffe (2021).

10 See Isaacson (2008) and Stachel (2002). The quote is from Hume (1739).

11 Isaacson (2008).

12 See Esterson and Cassidy (2020) for an analysis of Marić's role during this period of Einstein's life.

13 Isaacson (2008).

14 Ibid.

15 Ibid.

16 The non-variability of the speed of light was confirmed experimentally by US physicists Albert A. Michelson and Edward Morley in 1887. They tried to measure how the speed of light changed in relation to the direction in which the Earth was moving, but could determine no such variation. Their result was a blow to theories that postulated the existence of a special medium known as the 'ether', through which light was assumed to travel.

17 It's the person who has travelled to Proxima Centauri who has aged less, because that person had to turn around at the star and travel back to Earth. The person's travelling state was thereby altered, unlike that of the person who remained on Earth. If the asymmetry between the two people is not taken into account, there is a risk of drawing contradictory conclusions from this course of events, which is generally referred to as the twin paradox.

18 Neffe (2021).

19 Isaacson (2008).

20 Minkowski (1909).

CHAPTER 4

1 Gleick (2007).

2 Newton (1687).

3 Miller (2002).

4 Einstein (1922).

5 This does not apply to laboratories that are large enough so that gradients in a gravitational field can be experimentally measured.

6 Isaacson (2008).

7 Interview with Einstein in the *Saturday Evening Post*, 26 October 1929.

8 Isaacson (2008).

9 https://www.smithsonianmag.com/smart-news/ancient-tablet-shows-babylonians-used-pythagorean-geometry-1000-years-pythagoras-180978376/

10 My description of non-Euclidean geometry largely follows Bardi (2008), Richards (2002) and Cowen (2019).

11 Nethington (2020).

12 https://www.bbc.co.uk/blogs/waleshistory/2010/06/robert_recorde.html

13 This and other quotes are from Bardi (2008).

14 Bardi (2008).

15 Levenson (2004).

16 Einstein (2005).

17 https://einsteinpapers.press.princeton.edu/vol8a-doc/242

18 Van Dongen (2013).

19 Letter to Arnold Sommerfeld, 28 November 1915, cited in van Dongen (2013).

20 See Einstein (1915a).

21 In popular science contexts, spacetime curvature is often presented as a surface that gets pressed down. This picture is somewhat misleading when it comes to objects such as Earth, because it is primarily the curvature of time that influences how things move towards and around Earth. The curvature of time is, however, considerably harder to visualize.

22 Thorne (2014).

23 https://www.nist.gov/news-events/news/2010/09/nist-clock-experiment-demonstrates-your-head-older-your-feet

24 Isaacson (2008).

25 Levenson (2016).

26 See Einstein (1915b).

27 Einstein presented his calculation concerning Mercury's perihelion to the Academy of Sciences on 18 November 1915. The Academy's records show that Schwarzschild delivered his artillery fire report on the same date. The report was confidential and was not made public until after the war; see Schwarzschild (1920). There are no letters between Schwarzschild and his wife Else for this period, which is most likely due to Schwarzschild being in Berlin, from where he would have been able to meet her easily at their home in Potsdam just outside the city (or he might have stayed in Potsdam and visited the meeting at the Academy). If Schwarzschild had heard Einstein's presentation in person, this would explain why he was able, so quickly, to make a start on his own calculations for Mercury's perihelion.

CHAPTER 5

1 Letter to Else Schwarzschild, 3 December 1915.

2 https://einsteinpapers.press.princeton.edu/vol8a-doc/296

3 https://einsteinpapers.press.princeton.edu/vol8a-doc/303

4 Rindler (1956).

5 Gray (2009).

6 https://www.nobelprize.org/prizes/physics/2020/advanced-information/

7 The extreme time delay would only be possible if the planet's orbit was extremely close to the black hole's event horizon. The planet would be travelling at almost the speed of light. The scenario is not realistic, but it works well as an illustration of the relativity of time. See Thorne and Nolan (2014) and Luminet (2015).

8 Finkelstein (1958). There is also a mathematical solution for *white holes*, a sort of inverted black hole. Everything in a white hole must travel from its inside to its outside.

NOTES

323

Travelling into a white hole is impossible. Unlike black holes, however, most physicists believe it is impossible for white holes to exist, but see Rovelli (2023) for a fascinating description of the role white holes could play in the universe.

9 Curiel (2019).

10 The Schwarzschild metric also produces a singularity at the event horizon. This singularity created serious headaches for Einstein and others, but a detailed mathematical analysis showed that it resulted from the coordinates the formula was written in, and therefore had no physical significance. The Dutch physicist Johannes Droste devised the modern way of writing the formula, without a coordinate restriction that Schwarzschild imposed, in his doctoral thesis in 1916.

11 The forces that stretch your body in one direction and squeeze it in another are called tidal forces. The name is derived from the effect of the Moon and the Sun's gravitational fields on the Earth's oceans, namely, the tides. They arise because of the variation in the strength of the gravitational force from the Moon and the Sun across the size of the Earth. Similarly, the variation in the gravitational strength around and inside a black hole can deform a body that is orbiting or falling into it.

12 Schwarzschild (1916a) and Schwarzschild (1916b).

13 A common assertion in black hole literature is that Schwarzschild picked up the skin condition when he was on the Eastern Front (and that he discovered the formula for black holes there too). While it is true that Schwarzschild did spend a short period of 1915 on a military assignment in Lithuania, his illness was probably genetic, and not due to the war.

14 Einstein (1916).

15 Letter to Michele Besso, 14 May 1916. https://einsteinpapers.press.princeton.edu/vol8a-doc/358

16 Einstein (1939).

CHAPTER 6

1 https://www.eso.org/public/news/eso2408/

2 BH stands for 'black hole', and the number refers to the order in which the discoveries were presented. Gaia BH2 is 3,800 light years away and has a mass of almost nine solar masses.

3 Goethe, J.W., *Faust*, The World Publishing Company, 1870, tr. Bayard Taylor.

4 Stephenson and Green (2003).

5 The name was introduced by the Irish astronomer William Parsons. In the mid-nineteenth century he drew a picture of the nebula and thought it looked like a crab.

6 Chadwick was awarded the 1935 Nobel Prize in Physics for this discovery.

7 See the two references to Baade and Zwicky from 1934 in the Sources section. Similar ideas were presented by the physicist Lev Landau. In 1987 a supernova exploded in the Large Magellanic Cloud, which was an important event for astronomers. For the first time they were able to study the details of this kind of explosion close at hand. In 2024 the supernova made headlines around the world when a group of astronomers, including Claes Fransson from Stockholm University and Josefin Larsson from KTH Royal Institute of Technology, showed that the supernova had led to a neutron star rather than a black hole. See Fransson et al. (2024).

8 This statement was made on 11 January 1935 at a meeting of the Royal Astronomical Society.

9 Israel (1996).

10 Oppenheimer and Volkov (1939), Tolman (1939).

11 The same day Hitler's troops invaded Poland, Wheeler had published an article with the Danish physicist Nils Bohr, in the same journal as Oppenheimer and Snyder. In the article, the fission process in the atomic nucleus, that is, the splitting of the atom, was analysed.

12 Wheeler (1968).

13 White dwarfs can also explode in supernovas if they pull in gas from a star partner and approach the Chandrasekhar limit of 1.4 solar masses. However, the explosion would tear the whole star apart, leaving no neutron star or black hole behind.

14 Gold (1969).

15 https://www.aip.org/history-programs/niels-bohr-library/oral-histories/31792

16 https://www.gla.ac.uk/explore/avenue/me/mebyjocelynbellburnell/

17 As well as a three-hour interview with Roy Kerr, I have found valuable information about his life in Kerr (2008) and Melia (2009).

18 Thorne (1974), see also Baarden (1973).

19 Several physicists have studied whether there might be other characteristics that would make two black holes of the same mass, rotation and electrical charge different to one another. They demonstrated that it was impossible, which tends to be referred to by the phrase 'black holes have no hair', that is, there are no microscopic properties around them that would distinguish them.

20 Melia (2009).

21 It was for studies of Cygnus X-1 that the first Nobel Prize awarded for work on black holes was given. The US-Italian astronomer Riccardo Giacconi was awarded the prize in 2002 for his groundbreaking analysis of X-rays from space. In the late seventies the X-ray telescope Uhuru was launched from a space base off the coast of Kenya. One of the telescope's primary aims was to study the X-rays emitted from Cygnus X-1 and confirm whether or not they came from a black hole. It was with the help of Uhuru that astronomers were able to estimate the size of the dark object in Cygnus X-1.

22 Zhao et al. (2021).

CHAPTER 7

1 As well as visits to LIGO Hanford and interviews, I have based this chapter on information from Thorne (1994), Bartusiak (2000), Kennefick (2007), Maggiore (2007), Levin (2017), Collins (2018), Maggiore (2018) and Schilling (2019).

Besides my interview with Weiss and the books above, I have found https://www.kavliprize.org/rainer-weiss-autobiography very useful.

2 The piano's unique sound is due to the fact that each note consists not only of one kind of oscillation, but several. The combination of different kinds of oscillations, called overtones, gives every instrument its particular sound. Among these is a fundamental base frequency that is stronger than the others, and it is this frequency that determines the pitch.

3 *The construction, operation, and supporting research and development of a Laser Interferometer Gravitational-wave Observatory* by Ronald W.P. Drever, Kip S. Thorne, Frederick J. Raab and Rainer Weiss. The quote from Machiavelli's *The Prince* is from W.K. Marriott's translation.

4 One of the black holes was thirty-six solar masses, the other twenty-nine. After the merger, a black hole of sixty-two solar masses was created. The difference of three solar masses had been lost through the production of gravitational waves.

5 Abbott et al. (2017).

6 LIGO's analysis of the signal went all the way up to 2,000 hertz, but most of the information in the signal was at lower frequencies up to around 500 hertz.

7 There are also plans to construct gravitational wave detectors in India (called LIGO India) and in China.

8 A similar project is being planned in China, under the name TianQin.

NOTES

325

CHAPTER 8

1 This chapter is based on conversations with several members of the EHT and visits to a number of telescopes: Submillimeter Array in the US, IRAM 30-meter in Spain, South Pole Telescope in the Antarctic and NOEMA in France. Falcke, Melia and Agol (2020) and Falcke (2022) provide a first-hand account of Event Horizon Telescope's work. For a detailed description of the work that led to the observations in 2017, see Fletcher (2018). An excellent survey of the work involved in producing the first image of a black hole can be found in Skulberg (2021).

The description is based on an interview with Luminet. See also Skulberg and Sparre (2023). A survey of the history of black hole imaging can be found on Luminet's blog: https://blogs.futura-sciences.com/e-luminet/2018/03/07/45-years-black-hole-imaging-1-early-work-1972-1988/.

2 Gérard de Nerval's poem was part of the sonnet sequence *Les Chimères*, published in *Les Filles du feu*, and was translated into English by Jean-Pierre Luminet.

3 Carter and Luminet (1978). The French article published in 1978 contained an overview of the new results. In January 1979, Luminet published a scientific article in English with all equations and diagrams in the well-known international journal *Astronomy and Astrophysics*.

4 Luminet (1979).

5 Two meetings took place, one after the other, at Parque de las Ciencias. First, the members of Event Horizon Telescope had an invitation-only meeting. Then there was an open meeting for Next Generation Event Horizon Telescope that I took part in.

6 https://www.voxweb.nl/english/the-rising-star-of-sara-issaoun

7 The first calculation of how light curves around a rotating black hole was undertaken by Godfrey (1970). After that, Baarden (1973) and Cunningham and Bardeen (1973) studied how the light from a starfield and a single star, respectively, would be influenced by the black hole's gravity. They made their calculations with pen and paper. Luminet's (1979) calculation was the first calculation done on a computer of the effect on light from the gas around a black hole. He still had to draw his final results by hand, using the computer calculations as a basis. Fukue and Yokoyama (1988) produced the first colour depictions of how the intensity and wavelength of light around a black hole change. During the nineties, improvements in computer capacity and graphics programs led to more researchers exploring ways to visualize the surroundings of a black hole. The French astrophysicist Jean-Alain Marck, who worked with Luminet, made a film in 1991 in which the viewer flies through the gas close to a black hole. Viergutz (1993) carried out advanced computer calculations and visualizations of the effect on the light from a disc of gas around a rotating black hole. The 2000 article by Falcke, Algol and Melia was groundbreaking, because it showed how the shadow of a black hole would look to a telescope on the Earth, thereby connecting the simulated visualizations to potential observations using VLBI.

8 Around this time, David Hilbert and Max von Laue also conducted important calculations relating to how light trajectories bend around a celestial body.

9 See Kennefick (2019) for an entertaining portrayal of how the expedition and photographic analysis took place.

10 The observations were taken at a wavelength of 1.3 millimetres.

11 https://www.aip.org/history-programs/niels-bohr-library/oral-histories/46822

12 Doeleman et al. (2008) and Doeleman et al. (2012).

13 Lu et al. (2014).

14 Atomic clocks, hard drives, new software and other technical improvements were required to make a number of the telescopes ready for the Event Horizon Telescope observations. ALMA, which consisted of several linked telescopes in the Atacama desert, required an especially extensive overhaul in order that its data could be utilized. Doeleman was instrumental in the work required to make ALMA a functional part of the network.

15 The following telescopes contributed to the 2017 observations. USA: Submillimeter Telescope, James Clerk Maxwell Telescope, Submillimeter Array. Chile: Atacama Pathfinder Experiment, Atacama Large Millimeter/Submillimeter Array. Spain: IRAM 30-meter Telescope. Mexico: Large Millimeter Telescope Alfonso Serrano. Antarctic: South Pole Telescope.

16 Skulberg (2021).

17 From the documentary *Black Holes: The Edge of All We Know*.

18 https://www.nature.com/articles/nature.2015.16830

19 https://www.nytimes.com/2014/03/18/science/space/detection-of-waves-in-space-buttresses-landmark-theory-of-big-bang.html

20 In the late sixties the Swedish radio astronomer Jan Högbom developed an algorithm that two of the groups used. Högbom began his career as an astronomer at the Stockholm Observatory. After gathering data using radio telescopes on the east coast of America, he tried to turn this data into images, but saw only 'irrelevant details' and 'junk'. He therefore developed his own algorithm to extract an image from the incomplete data. Högbom called this algorithm CLEAN and presented it in a paper in 1974. The algorithm was highly effective and was used to create images of everything from underwater objects, space junk, subterranean structures and also black holes.

21 Event Horizon Telescope Collaboration (2019).

22 https://eventhorizontelescope.org/press-release-april-10-2019-astronomers-capture-first-image-black-hole. EHT had produced several images from the days of black hole observations. The image that was presented at the press conference was from 11 April 2017. The image differs from the black hole in *Interstellar* for several reasons. One is that the observation angle in *Interstellar* is different. Another is that the gas in *Interstellar* forms a thin disc, while around M87* it is more diffuse. *Interstellar* also uses high resolution computer animations, meaning that everything that happens around the black hole is much sharper than in the real image. In *Interstellar* the Doppler effect, which makes part of the gas around the black hole appear lighter and part darker, has also been omitted as it was deemed visually confusing.

23 Christensen et al. (2019).

24 Event Horizon Telescope Collaboration (2019).

25 https://hubblesite.org/contents/media/images/2000/20/968-Image.html

26 At this time I was working as a science communicator at the European Southern Observatory and was involved in producing a press release about the GMVA result. See https://eso.org/public/news/eso2305/.

27 The analysis investigated how the ring had altered between the 2017 and 2018 observations. By 2018, new telescopes had joined the EHT network: IRAM NOEMA Observatory, Kitt Peak Telescope and Greenland Telescope.

28 https://eventhorizontelescope.org/M87-one-year-later-proof-of-a-persistent-black-hole-shadow

29 https://www.eso.org/public/news/eso2406/

30 Medeiros et al. (2023).

31 The EHT managed to observe at a wavelength of 0.87 millimetres, compared to the previous observations that were made at 1.3 millimetres. See Raymond et al. (2024).

NOTES
327

CHAPTER 9

1 Volonteri (2010).
2 The basic elements of how supermassive black holes can be created were worked out by Marta Volonteri's former colleague Martin Rees, among others, in the late seventies.
3 Astronomers have also speculated that there could be black holes with masses of several thousand up to more than a hundred thousand Suns. These do not belong to the category 'supermassive black holes', but are instead classified as 'intermediate-mass black holes'. They could exist, for example, at the centre of globular star clusters consisting of hundreds of thousands of stars. In the summer of 2024 astronomers reported that they had identified one such black hole, with a mass around 8,200 times that of the Sun, in the star cluster Omega Centauri. Half a year later, another group of astronomers claimed that it was more likely that there was an extended object, formed by, for example, a cluster of stellar-mass black holes, at the centre of Omega Centauri. The science case of Omega Centauri highlights how difficult it is to pinpoint the existence of an intermediate mass black hole. See https://esahubble.org/news/heic2409/ and https://www.iac.es/en/outreach/news/iac-researchers-find-stars-omega-centauri-move-under-action-cluster-black-holes.
4 The British astronomer Arthur Eddington studied how the outward light pressure can be balanced against the inward pull of gravity in the context of stars. The maximum rate at which black holes can consume matter is known as the Eddington limit after his initial study.
5 Gillispie (2000).
6 Schmidt (1963).
7 In 1964 Edwin Salpeter in the US and Yakov Zeldovich and Igor Novikov in the Soviet Union all independently suggested that supermassive black holes were what powered quasars. Since then, several models have been put forward, describing the astrophysical details. David Lynden-Bell and Martin Rees developed the models of how black holes might create the radiation of quasars. In the early seventies Roger Penrose showed how energy could be extracted from a black hole's rotation, and in 1977, Roger Blandford and Roman Znajek presented a detailed model that showed how magnetic fields and the gas in an accretion disc can channel the energy from this rotation into jets.
8 https://www.eso.org/public/news/eso2402/
9 Fan et al. (2004).
10 Zeldovich and Novikov (1967). They asserted that primordial black holes could grow so big that if they existed, they ought to have been observed, and therefore assumed that they did not exist. Hawking (1971) and Carr and Hawking (1974) demonstrated that this conclusion was wrong.
11 Hawking (1971).
12 https://www.nature.com/articles/nature.2012.10765. A study published in June 2025 showed that the probability of a collision between Milky Way and Andromeda is much lower than previously estimated. See https://www.helsinki.fi/en/news/space/no-certainty-about-predicted-milky-way-andromeda-collision
13 Schiavi (2020).
14 Lodato and Natarajan (2006).
15 Natarajan et al. (2017).
16 This result was also established with the Hubble Space Telescope, see Hayes et al. (2024).
17 See Maiolino et al. (2024).
18 See Übler et al. (2024) and Suh et al. (2025).
19 Castellano et al. (2023).

20 The research institute is run by Harvard College Observatory and Smithsonian Astrophysical Observatory.
21 Bogdán et al. (2024) and Natarajan et al. (2024).
22 Together with other astronomers, Priya Natarajan and Marta Volonteri identified a further object similar to UHZ-1, called GHZ9. See Kovács et al. (2024).

CHAPTER 10

1 Marichalar (2021).
2 The name's contested status is described in Salazar (2014).
3 https://theconversation.com/mauna-a-wakea-hawaiis-sacred-mountain-and-the-contentious-thirty-meter-telescope-46069
4 https://observers.france24.com/en/20141015-hawaiians-telescope-sacred-volcano-protest
5 https://18millionrising.org/2019/11/Maunakea.html
6 https://www.hawaiipublicradio.org/general-assignment/2015-05-15/how-the-debate-over-tmt-prompted-a-problematic-email
7 https://www.ucolick.org/statement-faber.html
8 For a timeline, see https://kawaiola.news/aina/mauna-kea-timeline/.
9 McCormmach (2012), Tyson and Lang (2018) and https://www.smh.com.au/national/nsw/the-timepiece-that-helped-captain-cook-get-a-quick-fix-and-put-australia-on-the-map-20160427-gogacn.html.
10 Sparks (1847).
11 The loss of land was a consequence of the introduction of a private system of land ownership in 1848. The goal was to guarantee land rights for Indigenous people, but in fact it had the opposite effect. See https://files.hawaii.gov/dcca/reb/real_ed/re_ed/ce_prelic/land_in_hawaii.pdf
12 https://harvardlawreview.org/print/vol-133/aloha-aina-native-hawaiian-land-restitution
13 https://spectrum.ieee.org/edison-and-the-king-how-hawaii-became-electrified
14 A hundred years after the 1893 coup d'état, US President Bill Clinton signed an 'Apology Resolution', apologizing for the removal of the Indigenous people's right to self-determination.
15 https://www.focusongeography.org/publications/articles/toma/index.html
16 Stanley (2020).
17 The astronomer Karl Schwarzschild also took part in a 1905 expedition to Algeria to observe a solar eclipse. The observations were made possible by a grant of permission from the French colonial powers and the now-forgotten labour of the local Algerian people.
18 See Wali (1991), Wali (1997) and Miller (2005).
19 https://inthesetimes.com/article/cold-war-militarism-greenland-inuit-arctic
20 The details in this paragraph are taken from the European Court of Human Rights document HINGITAQ 53 vs. Denmark, Application No. 18584/04. See https://www.elaw.org/content/denmark-hingitaq-53-vs-denmark-application-no-1858404-decision-european-court-human-rights-a.
21 https://heasarc.gsfc.nasa.gov/docs/xlcalibur/
22 https://www.sametinget.se/1054
23 Sörlin and Wormbs (2010), Backman (2015).
24 https://www.esa.int/esapub/bulletin/bullet88/peder88.htm. See also Sheehan (2018) and Stiernstedt (2001).
25 In Backman (2009), it is summarized thus: 'The Sámi were seen as part of the natural environment without that implying the right to the ownership of it. The state viewed the territory as ownerless, terra nullius, and therefore felt entitled to claim it as theirs.'

NOTES
329

26 https://www.svt.se/nyheter/lokalt/norrbotten/esrange-utvecklingsplaner-hotas
27 https://nsd.se/nyheter/kiruna/artikel/talma-och-esrange-overens-om-utbyggnad/1108280j
28 https://www.svt.se/nyheter/sapmi/esrange-vill-skjuta-upp-satelliter-samebyar-kritiska-vara-marker
29 https://air-cosmos.com/article/il-y-a-50-ans-la-france-quittait-la-base-dhammaguir-en-algrie-4521
30 https://foreignlegion.info/history/3rei/
31 https://www.la-croix.com/France/Politique/Guyane-Centre-spatial-guyanais-occupe-manifestants-2017-04-05-1200837341
32 https://qz.com/960817/how-a-handful-of-south-american-protestors-in-french-guiana-took-arianespace-and-europes-space-program-hostage
33 https://www.france24.com/en/20170422-french-guiana-protests-end-with-agreement-promise-aid
34 https://www.northropgrumman.com/space/james-webb-space-telescope
35 https://www.lockheedmartin.com/en-us/news/features/history/hubble.html
36 Tyson and Lang (2018).
37 Cited in Tyson and Lang (2018).
38 For further examples, see Tyson and Lang (2018).
39 Melia (2009).
40 The event is described in Chatterjee (2012). See also Bartusiak (2016), and Herdeiro and Lemos (2019).
41 Wheeler (2000).
42 Bartusiak (2016) and Thorne (1994). This obscene metaphor wasn't improved by Wheeler going on to coin the phrase 'black holes have no hair' to describe how information about what has fallen into a black hole disappears, leaving nothing but a smooth surface, that is, 'no hair'.
43 https://www.bbc.com/future/article/20200803-the-forgotten-mine-that-built-the-atomic-bomb
44 This took place in 1947. The inhabitants were relocated to the Ujelang atoll, around 200 kilometres from their original home. See https://www.dtra.mil/Portals/61/Documents/NTPR/1980-DNA Fact Sheet_Enewetak Operation.pdf.
45 https://fluxhawaii.com/written-in-the-stars/
46 https://en.wal.unesco.org/
47 https://www.nationalgeographic.com/culture/article/hawaii-native-language-efforts-to-preserve-revive
48 https://imiloahawaii.org/a-hua-he-inoa
49 https://www.eaobservatory.org/jcmt/2019/04/powehi/
50 https://www.capjournal.org/issues/26/26_11.pdf
51 https://www.kona-kohala.com/news/details/powehi-day-proclamation
52 About a year after the naming of Pōwehi, NASA declared it would no longer call the planetary nebula NGC 2392 the 'Eskimo nebula'. The nebula was given the name because it was thought to resemble an Inuit anorak. In a press release, NASA wrote that the word '"eskimo" is widely viewed as a colonial term with a racist history'. NASA would also stop using the nickname 'Siamese Twins Galaxy' for the colliding galaxies NGC 4567 and NGC 4568. The phrase 'Siamese twin' originates from two conjoined twins from Thailand (formerly referred to as Siam) who were paraded at circuses and in other contexts in the USA in the 1830s. See https://www.nasa.gov/feature/nasa-to-reexamine-nicknames-for-cosmic-objects.
53 https://fluxhawaii.com/written-in-the-stars/
54 https://www.democracynow.org/2019/7/22/why_indigenous_protectors_oppose_the_thirty

330   FACING INFINITY

55 See Cuby et al. (2024) for background information regarding the Maunakea Stewardship and Oversight Authority and its development.
56 An external evaluation committee stated towards the end of 2024 that the National Science Foundation had to choose to fund one of two potential extremely large US telescopes: the TMT on Hawai'i (with an alternative site on La Palma, Canary Islands) or the Giant Magellan Telescope in Chile.
57 As part of Next Generation Event Horizon Telescope, a working group has been established to promote adherence to ethical guidelines in the construction of new telescopes.

CHAPTER 11

1 https://www.nrdc.org/bio/rob-moore/ipcc-report-sea-level-rise-present-and-future-danger
2 *The Odyssey of Homer*, tr. Lattimore (1965), Harper & Row.
3 https://www.phaidon.com/agenda/design/articles/2017/october/26/how-buzz-aldrin-saw-the-night-sky/
4 https://earthdata.nasa.gov/learn/sensing-our-planet/beacons-in-the-sky-help-monitor-earth-s-orientation-in-space
5 Carter and Robertson (1986).
6 https://ivscc.gsfc.nasa.gov/about/vlbi/NASA-GSFC_Geodetic_VLBI_Program.pdf
7 McCormmach (2012).
8 Raap (2024). As for Chapter 3, I thank Raap for sharing with me his materials concerning Karl Schwarzschild and for providing important information about his life.
9 Letter to Else Schwarzschild, 3 January 1916.
10 Unless otherwise stated, these quotes from Wegener and the descriptions of his life are taken from Yount (2009).
11 Letter to Else Schwarzschild, 3 May 1915.
12 Carlotto (2022).
13 Hapgood (1958).
14 Israel (1996).
15 Another interesting explanation for why so many astronomers and physicists doubted the possibility of black holes has been suggested by the Costa Rican astrophysicist M. Ortega-Rodríguez and his colleagues. According to them, the doubt was due not only to the extreme physical conditions. There was also an overarching intellectual model that characterized several scientific disciplines in the thirties: the idea of equilibrium. Several areas were dominated by the thought that a system moves towards equilibrium. Within economics, for instance, there was the idea of market equilibrium; in biology, biological equilibrium. This idea also dominated Einstein's early (and incorrect) cosmological models, in which he assumed that the universe was in equilibrium, and that it was neither expanding nor collapsing. The idea that a star might collapse indefinitely went against the idea of equilibrium, and therefore against a prevailing model for conceiving of the world at that time. See Ortega-Rodríguez et al. (2017).
16 Stewart (1990).
17 Yount (2009).
18 https://arstechnica.com/science/2014/05/how-el-nino-temporarily-slowa-the-earths-rotation/
19 https://astronomynow.com/2020/03/12/fossil-shows-days-were-a-half-hour-shorter-during-age-of-dinosaurs/
20 https://www.forbes.com/sites/jamiecartereurope/2023/07/13/60-hour-days-prevented-only-by-billion-year-pause-say-scientists/
21 See Shahvandi et al. (2024) and Agnew (2024).

NOTES   **331**

22 The time standard based on the Earth's rotation is called UT1, while UTC is the standard based on several atomic clocks.
23 Haas (2021).
24 https://science.nasa.gov/learn/basics-of-space-flight/chapter13-1/
25 https://www.cosmos.esa.int/web/gaia/edr3-acceleration-solar-system
26 https://www.ipcc.ch/report/ar6/wg2/chapter/chapter-3/

CHAPTER 12

1 The documentary *Leben aus dem All: Schwarze Löcher.*
2 https://www.nobelprize.org/prizes/physics/1967/bethe/lecture/
3 https://www.nasa.gov/universe/suzaku-shows-clearest-picture-yet-of-perseus-galaxy-cluster/
4 https://www.nasa.gov/universe/new-nasa-black-hole-sonifications-with-a-remix/
5 Simionescu et al. (2015). Similar observations were conducted using the X-ray telescope XMM-NEWTON; see Simionescu et al. (2008).
6 See Oei et al. (2024) and https://www.caltech.edu/about/news/gargantuan-black-hole-jets-are-biggest-seen-yet.
7 Gloudemans et al. (2025).
8 See Scharf (2013) for more information on the study of black holes through X-rays and their link to the origins of life. For a fascinating and highly speculative account of how supernovae and black holes could have influenced the chirality of the amino acids that form the building blocks of life; see Boyd (2012).
9 https://www.esa.int/Science_Exploration/Space_Science/Brightest_gamma-ray_burst_illuminates_our_galaxy_as_never_before
10 Hayes and Gallagher (2022).
11 https://www.indiatoday.in/science/story/earth-was-hit-by-a-massive-burst-of-energy-it-came-from-outside-solar-system-2463009-2023-11-15
12 https://news.northwestern.edu/stories/2024/04/brightest-gamma-ray-burst-of-all-time-came-from-the-collapse-of-a-massive-star/
13 Melott et al. (2004), Thomas and Melott (2006).
14 Subrayan et al. (2023).
15 Gehrels et al. (2002).
16 Garofolo (2023).
17 Balbi and Tombesi (2017).
18 Kendall et al. (2025).
19 Wada et al. (2021); see also Wada et al. (2019).
20 Opatrný, Richterek and Bakala (2017), Bakala, Dočekal and Turoňová (2020).
21 Schnittmann (2019).

CHAPTER 13

1 Ferguson (2019).
2 My Hawking story is based on several sources, including Hawking (1993), Hawking (1998), Hawking (2014), White and Gribbin (2016), Ferguson (2019), Mlodinow (2020), Seife (2021) and the Twinch documentary *Hawking: Can You Hear Me?*
3 Thorne described this trip in Thorne (1994).
4 This argument was based on Roger Penrose's analysis of the energy that can be extracted from black holes.
5 Hawking's calculation differs from this popular science version. He did not analyse these virtual particles close to the event horizon, but rather the properties of a vacuum in a quantum field around the black hole. The math was abstract and hard to follow,

and so Hawking presented his calculations in terms of the virtual particles in popular science contexts.

6 A black hole does not glow like a lump of coal. A glowing piece of coal gives off light. Black holes do also radiate light, but they give off all the other possible particles that exist in nature, such as neutrinos and electrons. A lump of coal grows cooler and cooler the longer it radiates its heat away. Its temperature decreases. A black hole works the other way round: its temperature increases the more it radiates. An additional difference is that the coal's heat radiation is created by the random movements of all the atoms and molecules that make up the lump. For a black hole, the radiation is created by the very structure of spacetime, and is not completely localized at the event horizon. In spite of these differences, the radiation from the coal and the black hole have one thing in common. Hawking showed that the energy distribution of the black hole's radiation has the same structure as the heat radiation of a glowing piece of coal, which is known as black body radiation.

7 Hawking (1974).

8 Rees (1977).

9 O'Sullivan (2018).

10 https://www.smartcompany.com.au/startupsmart/advice/startupsmart-growth/csiro-scientists-nominated-for-european-inventor-award/

11 https://www.npr.org/2018/03/27/597390626/an-engineers-quest-to-save-stephen-hawkings-voice

12 https://www.telegraph.co.uk/news/features/11609068/Jane-Hawking-Living-with-Stephen-made-me-suicidal-but-I-still-love-him.html

13 Esterson and Cassidy (2020).

14 Some interpretations of quantum mechanics maintain the existence of a non-determinism in nature. This illustrates how the question of predictability occurs both in the quantum world of particles and the gravitational world of black holes.

15 See Susskind (2008) for an entertaining and authoritative survey of the information paradox and the holographic principle.

16 Heino Falcke and his colleagues have developed a model in which all objects, not only black holes, give out the equivalent of Hawking radiation. Perhaps this means that over incredibly long timescales, objects such as planets and stars will also disappear, changing space and time around them. See Wondrak, van Suijlekom and Falcke (2023).

17 To begin with, Hawking himself believed that information could disappear forever. He was so convinced of this that he made a bet to that effect with Kip Thorne and John Preskill in 1991. Thirteen years later, Hawking admitted defeat. Via a complicated mathematical argument, he asserted that black holes' event horizons did not form fully, and that information could therefore not be fully lost inside a black hole. But Thorne refused to give up, and several other physicists still subscribe to Hawking's original conviction. See https://www.quantamagazine.org/stephen-hawkings-black-hole-paradox-keeps-physicists-puzzled-20180314/.

18 This was formulated by Leonard Susskind, Larus Thorlacius and Gerard 't Hooft. See Susskind (2008).

19 https://www.dailymail.co.uk/sciencetech/article-1052354/Are-going-die-Wednesday.html

20 https://hotnews.ro/update-small-romanian-party-sparks-mockery-saying-lhc-experiment-may-create-tiny-black-holes-and-that-cern-experiment-should-be-halted-775317

21 Giddings and Thomas (2002).

22 https://www.nytimes.com/2008/03/29/science/29collider.html

23 https://home.cern/science/accelerators/large-hadron-collider/safety-lhc

NOTES

24 https://www.prospectmagazine.co.uk/culture/37375/the-mind-of-god-the-problem-with-deifying-stephen-hawking
25 Jackson and Ryan Jun (1973).
26 Beasly and Tinsley (1974).
27 Dai and Stojkovic (2024).
28 Scherrer (2025).
29 See Scholtz and Unwin (2020) and Siraj and Loeb (2020).
30 https://www.thegazelle.org/issue/132/in-memory-of-stephen-hawking
31 https://www.bbc.com/news/technology-30290540
32 https://www.nytimes.com/2021/04/18/books/review/hawking-hawking-charles-seife.html
33 https://www.prospectmagazine.co.uk/culture/37375/the-mind-of-god-the-problem-with-deifying-stephen-hawking
34 https://www.westminster-abbey.org/abbey-commemorations/commemorations/stephen-hawking
35 Several news sites reported on the message, for example https://edition.cnn.com/2018/06/14/europe/stephen-hawking-voice-black-hole-trnd/index.html.
36 The double system, which consists of a star and a black hole, is also known as 1A 0620-00.

CHAPTER 14

1 The letter written by André Larrue is in the collection of the Hartmannswillerkopf Franco-German Great War Historial.
2 Einstein and Rosen (1935). One of their aims was to construct singularity-free models of particles using spacetime geometries.
3 Thorne had begun to study the mathematical structure of wormholes more closely after his friend Carl Sagan, who was then working on his book *Contact*, asked Thorne if there was a possibility black holes could be portals to other stars. The producer Lynda Obst worked both on the film based on Sagan's book, which starred Jodie Foster, and *Interstellar*, which starred Matthew McConaughey (who also appeared in *Contact*). It could be said that through her work on producing these two Hollywood films, she has played a major role in shaping pop cultural understandings of wormholes.
4 In 2013 Leonard Susskind and Juan Maldacena posited an idea that links wormholes to the phenomenon quantum entanglement, seeking to solve the information paradox via theoretical means. See Maldacena and Susskind (2013).
5 See Morris and Thorne (1988) and Event Horizon Telescope Collaboration (2022).
6 See, for example, Easson and Brandenberger (2001), Hamilton (2013), Poplawski (2021), Brandenberger, Heisenberg and Robnik (2021), and Stuckey (1994).
7 Frolov, Markov and Mukhanov (1989).
8 Smolin (1992) and Smolin (1994).
9 Smolin claims that his idea actually makes predictions about various astrophysical phenomena, such as the maximum mass of neutron stars or the properties of the fields that drove the universe's expansion during its very first phase.
10 https://www.independent.co.uk/space/nasa-jwst-black-hole-multiverse-universe-b2717672.html
11 Shamir (2025).
12 See, for example, Melia (2003), Pathria (1972) and Artemova and Novikov (2002).
13 See Bartusiak (2010) for a fascinating depiction of this astronomical discovery.
14 Einstein added a cosmological constant to his theory, which had a repulsive gravitational effect. His goal was that this effect would counteract the pull of all the matter, giving rise to a static universe. When he realized that the universe was

actually expanding, he excluded the term, but today the cosmological constant is one of the most common explanations for dark energy. Its precise value causes theoretical problems, however, since it requires a high degree of fine-tuning in order to agree with the quantum field theory understanding of the properties of a vacuum. As to whether Einstein really used the phrase 'biggest blunder', see O'Raifeartaigh and Mitton (2018).

15 Strictly speaking, the observable universe is defined from the time light could have been travelling since the Big Bang, while the cosmic background radiation was created when the universe was 380,000 years old. In practice, this time delay makes little difference to the current size of the observable universe.

16 Based on 2018 data from the Planck satellite, see Planck Collaboration (2020) and https://www.mpg.de/7044245/Planck_cmb_universe.

17 Lineweaver and Patel (2023).

18 Since there is no convincing evidence that the universe exhibits any large-scale rotation, we can use Schwarzschild's formula for a non-rotating black hole.

19 Dark energy and dark matter are not the same thing. Dark matter is what causes cosmic structures to group together. Its gravitational force is attractive. Dark energy has the opposite effect. Its gravitational force is repulsive – it makes things move away from each other. Together, these energy forms constitute around 95 per cent of the total energy content of the universe.

20 Some galaxies are moving away from us faster than the speed of light. This is possible because Einstein's relativity theory says only that nothing can move faster than the speed of light in vacuum, while space itself can expand faster than that. This does not mean, however, that what happens in the regions expanding at faster than the speed of light is forever beyond our ability to observe them. See Davis and Lineweaver (2004).

21 Doran and Crawford (1999).

22 It is possible to show how the Schwarzschild relation can be derived from the equations for an expanding universe if it has a Euclidean geometry. The theoretical physicist Sean Carroll has written about this argument on his blog; see https://www.preposterousuniverse.com/blog/2010/04/28/the-universe-is-not-a-black-hole/.

23 The claim made in Shamir (2025) was based on an analysis of the rotation of 263 galaxies observed by the JWST. One would expect the rotation of the galaxies to be random, with one half of the sample rotating in one direction and the other half in the opposite direction. But of the galaxies in the sample, 105 rotated counterclockwise and 158 clockwise in relation to the Milky Way. One possibility, the author claimed, was that this asymmetry could be due to the universe itself rotating, which, in turn, could arise if the universe was embedded in a spinning black hole. The exact mechanism for how such an embedding would influence the galactic rotation properties was, however, lacking. For further arguments concerning the conclusions in Shamir (2025), see https://www.scientificamerican.com/article/do-we-live-inside-a-black-hole/.

24 Cited in Greene (2020); see also Adams (1997).

25 The quote is from a film shown at the Hartmannswillerkopf Great War Historial.

26 Duality also crops up in the mathematical description of black holes, in the form of AdS/CFT correspondence, which relates the properties of gravity to those of a special kind of quantum field. This mathematical duality is important in the analysis of the information paradox.

27 See https://news.northwestern.edu/stories/2025/02/flickers-and-flares-milky-ways-central-black-hole-constantly-bubbles-with-light/, https://newsroom.ucla.edu/releases/astrophysicists-capture-huge-gamma-ray-flare-supermassive-black-hole-m87, https://umbc.edu/stories/black-hole-jets-observed-forming-in-real-time/ and https://www.mpifr-bonn.mpg.de/pressreleases/2025/4.

SOURCES

INTERVIEWS

The following persons were interviewed either via email, in person or online and appear in the book (a longer list of persons supplying relevant background material can be found in the Acknowledgments): Geoff Bower, Michael Bremer, Avery Broderick, Jessica Dempsey, Shep Doeleman, Sheila Dwyer, Heino Falcke, Reinhard Genzel, Andrea Ghez, Rüdiger Haas, Sara Issaoun, Sam Keliiheleua, Roy Kerr, Larry Kimura, Michael Landry, Jean-Pierre Luminet, Priyamvada Natarajan, Aurora Simionescu, Kip Thorne, Marta Volonteri and Rainer Weiss.

DOCUMENTARIES

BBC Four, *How to See a Black Hole*, 2019
Galison, P., *Black Holes: The Edge of All We Know*, 2020
Kantara, J.A., *Leben aus dem All: Schwarze Löcher*, 2021
Krüger, S., *Quest for the Exact Position*, 2017
Twinch, O., *Hawking: Can You Hear Me?*, 2021

ARCHIVE COLLECTIONS

Material relating to Karl Schwarzschild comes from the Göttingen State and University Library.

BOOKS

Some of the popular science books in this list have not been referred to in the notes, but they are excellent sources for those keen to read more about black holes.
Al-Khalili, J., *Black Holes, Wormholes and Time Machines*, CRC Press, 2012
Backman, F., *Making Place for Space: A History of 'Space Town' Kiruna 1943–2000*, Umeå University, 2015
Bardi, J.S., *The Fifth Postulate*, Trade Paper Press, 2008
Bartusiak, M., *Einstein's Unfinished Symphony*, Joseph Henry Press, 2000
Bartusiak, M., *The Day We Found the Universe*, Knopf Doubleday, 2010
Bartusiak, M., *Black Hole*, Yale University Press, 2016
Beckwith, M.W., *The Kumulipo*, University of Hawaii Press, 2000
Blundell, K., *Black Holes: A Very Short Introduction*, Oxford University Press, 2015
Boyd, R.N., *Stardust, Supernovae and the Molecules of Life*, Springer, 2012
Burke, B.F., Graham-Smith, F., Wilkinson P.N., *An Introduction to Radio Astronomy*, Cambridge University Press, 2019
Chandrasekhar, S., *Truth and Beauty*, University of Chicago Press, 1990
Chatterjee, P., *The Black Hole of Empire*, Princeton University Press, 2012
Collins, H., *Gravity's Kiss*, The MIT Press, 2018

Cowen, R., *Gravity's Century*, Harvard University Press, 2019
Eddington, A., *The Internal Constitution of the Stars*, Cambridge University Press, 1926 (1988)
Eddington, A., *Stars and Atoms*, Yale University Press, 1927
Einstein, A., *Mein Weltbild*, Ullstein, 2005
Eisenstaedt, J., *The Curious History of Relativity*, Princeton University Press, 2007
Esterson, A., Cassidy, D. C., *Einstein's Wife*, The MIT Press, 2020
Falcke, H., *Licht im Dunkeln*, Klett-Clotta, 2020
Ferguson, K., *Stephen Hawking*, Pitkin Publishing, 2019
Fletcher, S., *Einstein's Shadow*, Ecco, 2018
Frolov, V.P., Novikov, I.D., *Black Hole Physics*, Springer Dordrecht, 1998
Galileo Galilei, *Sidereus Nuncius*, Thomas Baglioni, 1610
Galison, P., *Einstein's Clocks and Poincare's Maps*, W.W. Norton, 2003
Gilder, J., Gilder, A.-L., *Heavenly Intrigue*, Knopf Doubleday, 2005
Gillispie, C.C., *Pierre-Simon Laplace*, Princeton University Press, 2000
Gleick, J., *Isaac Newton*, Knopf Doubleday, 2007
Gooneratne, S., *The White Dwarf Affair*, ProQuest LLC, 2005
Gray, T., *The Schwarzschild Family*, Bound biographies, 2009
Greene, B., *Until the End of Time*, Penguin, 2020
Gubser, S.S., Pretorius, F., *The Little Book of Black Holes*, Princeton University Press, 2017
Gustafsson, B., *Svarta hål*, Fri Tanke, 2015
Hankins, T.L., *Science and the Enlightenment*, Cambridge University Press, 1985
Hapgood, C.H., *Earth's Shifting Crust*, Pantheon Books, 1958
Hawking, J., *Travelling to Infinity*, Alma Books, 2014
Hawking, S., *Black Holes and Baby Universes and Other Essays*, Bantam, 1993
Hawking, S., *A Brief History of Time*, Bantam, 1998
Hoffmann, D., *Einsteins Berlin: Auf den Spuren eines Genies*, Wiley-VCH, 2006
Holmes, R., *The Age of Wonder*, Pantheon, 2008
Hume, D., *A Treatise of Human Nature*, 1739
Impey, C., *Einstein's Monsters*, W.W. Norton, 2018
Isaacson, W., *Einstein: His Life and Universe*, Simon & Schuster, 2008
Joshi, P.S., *The Story of Collapsing Stars*, Oxford University Press, 2018
Kennefick, D., *Traveling at the Speed of Thought*, Princeton University Press, 2007
Kennefick, D., *No Shadow of a Doubt*, Princeton University Press, 2019
Laplace, P.-S., *Exposition du système du monde, tome second*, 1796
Levenson, T., *Einstein in Berlin*, Bantam, 2004
Levenson, T., *The Hunt for Vulcan*, Head of Zeus, 2016
Levin, J., *Black Hole Blues and Other Songs from Outer Space*, Vintage, 2017
Luminet, J.-P., *Black Holes*, Cambridge University Press, 1992
McCormmach, R., *Weighing the World: The Reverend John Michell of Thornhill*, Springer, 2012
Maggiore, M., *Gravitational Waves: Volume 1*, Oxford University Press, 2007
Maggiore, M., *Gravitational Waves: Volume 2*, Oxford University Press, 2018
Meier, D.L., *Black Hole Astrophysics*, Springer, 2012
Melia, F., *The Edge of Infinity*, Cambridge University Press, 2003
Melia, F., *Cracking the Einstein Code*, University of Chicago Press, 2009
Miller, A., *Albert Einstein's Special Theory of Relativity*, Springer New York, 1998
Miller, A., *Einstein, Picasso*, Basic Books, 2002
Miller, A., *Empire of the Stars*, Mariner Books, 2005
Mlodinow, L., *Stephen Hawking*, Pantheon, 2020
Moffat, J.W., *The Shadow of the Black Hole*, Oxford University Press, 2020
Moskvitch, K., *Neutron Stars*, Harvard University Press, 2020

SOURCES

Neffe, J., *Einstein: Eine Biographie*, Rowohlt Taschenbuch, 2021
Newton, I., *Philosophiæ Naturalis Principia Mathematica*, 1687
Pais, A., *Subtle Is the Lord: The Science and the Life of Albert Einstein*, Oxford University Press, 2005
Pascal, B., *Pensées sur la religion et sur quelques autres sujets*, 1670
Reinke-Kunze, C., *Alfred Wegener*, Springer, 1994
Rovelli, C., *White Holes*, Riverhead Books, 2023
Sanders, R.H., *Revealing the Heart of the Galaxy*, Cambridge University Press, 2014
Scharf, C.A., *Gravity's Engines*, Farrar, Straus and Giroux, 2013
Schilling, G., *Ripples in Spacetime*, Harvard University Press, 2019
Schwarzschild, K., *Gesammelte Werke Vol. 1*, Springer, 1992
Seife, C., *Hawking Hawking*, Basic Books, 2021
Skulberg, E., *The Event Horizon as a Vanishing Point*, University of Cambridge, 2021
Skulberg, E., *Invisibility as Argument: Visual Representations of Black Holes*, forthcoming
Smethurst, B., *A Brief History of Black Holes*, Macmillan, 2022
Sparks, J., *Life of John Ledyard, American Traveller*, Charles C. Little and James Brown, 1847
Stanley, M., *Einstein's War*, Penguin, 2020
Stewart, J.A., *Drifting Continents and Colliding Paradigms*, Indiana University Press, 1990
Stiernstedt, J., *Sweden in Space*, European Space Agency, 2001
Sullivan, W.T., *The Early Years of Radio Astronomy*, Cambridge University Press, 2005
Susskind, L., *The Black Hole War*, Little, Brown and Company, 2008
Tassoul, J.-L., Tassoul, M., *A Concise History of Solar and Stellar Physics*, Princeton University Press, 2014
Taylor, E.F., Wheeler, J.A., *Exploring Black Holes*, Addison Wesley Longman, 2000
Thorne, K., *Black Holes and Time Warps*, W.W. Norton, 1994
Thorne, K., Nolan, C., *The Science of Interstellar*, W.W. Norton, 2014
Tyson, N. deG., *Death by Black Hole*, W.W. Norton, 2007
Tyson, N. deG., Lang, A., *Accessory to War*, W.W. Norton, 2018
van Dongen, J., *Einstein's Unification*, Cambridge University Press, 2013
Venkataraman G., *Chandrasekhar and His Limits*, Universities Press (India), 1992
Verschuur, G., *The Invisible Universe*, Springer, 2007
Wali, K.C., *Chandra*, University of Chicago Press, 1991
Wali, K.C., *S. Chandrasekhar*, World Scientific, 1997
Wheeler, J.A., Ford, K., *Geons, Black Holes, and Quantum Foam*, W.W. Norton, 2000
White, M., Gribbin, J., *Stephen Hawking*, Pegasus Books, 2016
Yount, L., *Alfred Wegener*, Chelsea House, 2009

SCIENTIFIC ARTICLES, BOOK CHAPTERS AND ESSAYS

Abbott, B.P., et al., *Multi-messenger Observations of a Binary Neutron Star Merger*, Astrophysical Journal Letters, vol. 848, 2, 2017
Adams, F.C., *A dying universe: The long-term fate and evolution of astrophysical objects*, Reviews of Modern Physics, vol. 69, 1997
Agnew, D.C., *A global timekeeping problem postponed by global warming*, Nature, vol. 628, 2024
Artemova, I.V., Novikov, I.D., *The Interior of Black Holes and their Astrophysics*, arXiv: 0210545VI, 2002
Baade, W., Zwicky, F., *Cosmic Rays from Super-Novae*, Contributions from the Mount Wilson Observatory, vol. 3, 1934
Baade, W., Zwicky, F., *Remarks on Super-Novae and Cosmic Rays*, Phys. Rev., vol. 46, 76, 1934
Baarden, J., *Timelike and null geodesics in the Kerr metric*, Proceedings, Ecole d'Eté de Physique Théorique: Les Astres Occlus, 1972 (1973)

Backman, F., *Från föhn till feu! Esrange och den norrländska rymdverksamhetens tillkomsthistoria från sekelskiftet 1900 till 1966*, Umeå University, 2009

Bakala, P., Dočekal, J., Turoňová, Z., *Habitable Zones around Almost Extremely Spinning Black Holes (Black Sun Revisited)*, Astrophysical Journal, vol. 889, 1, 2020

Balbi, A., Tombesi, F., *The habitability of the Milky Way during the active phase of its central supermassive black hole*, Scientific Reports vol. 7, 1, 2017

Balick B., Brown R.L., *Intense sub-arcsecond structure in the galactic center*, Astrophysical Journal, vol. 194, 1974

Bardeen, J.M., *Kerr Metric Black Holes*, Nature, vol. 226, 1970

Beasly, W.H., Tinsley B.A., *Tungus event was not caused by a black hole*, Nature, vol. 250, 1974

Becklin, E.E., Neugebauer, G., *Infrared Observations of the Galactic Center*, Astrophysical Journal, vol. 151, 1968

Berggren, L., *Tycho Brahes mustasch postumt vittne till mord?*, Läkartidningen nr 46, vol. 106, 2009

Bogdán, A., et al., *Evidence for heavy seed origin of early supermassive black holes from a z ~ 10 X-ray quasar*, Nature Astronomy, vol. 8, 1, 2024

Brandenberger, R., Heisenberg, L., Robnik, J., *Through a Black Hole into a New Universe*, arXiv:2105.07166, 2021

Carlotto, M., *Toward a New Theory of Earth Crustal Displacement*, Journal of Scientific Exploration, vol. 36, 1, 2022

Carr, B.H., Hawking, S., *Black Holes in the Early Universe*, Monthly Notices of the Royal Astronomical Society, vol. 168, 2, 1974

Carter, B., 'Mathematical Foundations of the Theory of Relativistic Stellar and Black Hole Configurations' in *Gravitation in Astrophysics* (ed. Carter, B., Hartle, J.B.), Springer, 1987

Carter, B., Luminet, J.-P., *Les Trous Noirs: Maelströms Cosmiques*, La Recherche, no. 9, 1978

Carter, W.E., Robertson, D.S., *Studying the Earth by Very-Long-Baseline Interferometry*, Scientific American, vol. 255, 5, 1986

Castellano, M., et al., *Early Results from GLASS-JWST. XIX. A High Density of Bright Galaxies at z ~ 10 in the A2744 Region*, Astrophysical Journal Letters, vol. 948, 2, 2023

Charlot, P., et al., *The third realization of the International Celestial Reference Frame by very long baseline interferometry*, A&A, vol. 644, A159, 2020

Christensen, L.L., et al., *An Unprecedented Global Communications Campaign for the Event Horizon Telescope First Black Hole Image*, Communicating Astronomy with the Public Journal, vol. 26, 2019

Cuby, J.-G., et al., *Astronomy's relationship with the lands and communities of Maunakea*, Proc. SPIE, vol. 13094, 2024

Cunningham, C.T., Bardeen, J.M., *The Optical Appearance of a Star Orbiting an Extreme Kerr Black Hole*, Astrophysical Journal, vol. 183, 1973

Curiel, E., *The many definitions of a black hole*, Nature Astronomy, vol. 3, 2019

Dai, D.-C., Stojkovic, D., *Searching for small primordial black holes in planets, asteroids and here on Earth*, Physics of the Dark Universe, vol. 24, 2024

Davis, T.M., Lineweaver, C.H., *Expanding confusion: common misconceptions of cosmological horizons and the superluminal expansion of the Universe*, Publications of the Astronomical Society of Australia, vol. 21, 1, 2004

Doeleman, S., et al., *Event-horizon-scale structure in the supermassive black hole candidate at the Galactic Centre*, Nature, vol. 455, 78, 2008

Doeleman, S., et al., *Jet Launching Structure Resolved Near the Supermassive Black Hole in M87*, Science, vol. 338, 6105, 2012

Doran, R., Crawford, P., *Could the observable universe be inside of a black hole?*, Astrophysics and Space Science, vol. 263, 1999

SOURCES   **339**

Easson, D.A., Brandenberger, R.H., *Universe Generation from Black Hole Interiors*, JHEP, vol. 06, 024, 2001

Einstein, A., 'Die Feldgleichungen der Gravitation' in *Sitzungsberichte der Königlich Preußischen Akademie der Wissenschaften*, Verlag der Königlichen Akademie der Wissenschaften, 1915a

Einstein, A., 'Erklärung der Perihelbewegung des Merkur aus der allgemeinen Relativitätstheorie' in *Sitzungsberichte der Königlich Preußischen Akademie der Wissenschaften*, Verlag der Königlichen Akademie der Wissenschaften, 1915b

Einstein, A., 'Gedächtnisrede des Hrn. Einstein auf Karl Schwarzschild' in *Sitzungsberichte der Königlich Preußischen Akademie der Wissenschaften*, Verlag der Königlichen Akademie der Wissenschaften, 1916

Einstein, A., *How I created the theory of relativity*, 1922 [recounted in Physics Today, vol. 35, 8, 1982]

Einstein, A., *On a Stationary System With Spherical Symmetry Consisting of Many Gravitating Masses*, Annals of Mathematics, Second Series, vol. 40, 4, 1939

Einstein, A., Rosen, N., *The Particle Problem in the General Theory of Relativity*, Phys. Rev., vol. 48, 73, 1935

Eisenhauer, F., et al., *A Geometric Determination of the Distance to the Galactic Center*, Astrophysical Journal, vol. 597, 2, 2003

Eisenstaedt, J., *Histoire et singularités de la solution de Schwarzschild (1915–1923)*, Arch. Hist. Exact Sci., vol. 27, 1982

Eisenstaedt, J., *De l'influence de la gravitation sur la propagation de la lumière en théorie newtonienne. L'archéologie des trous noirs*, Arch. Hist. Exact Sci., vol. 42, 1991

Event Horizon Telescope Collaboration, *First M87 Event Horizon Telescope Results. I. The Shadow of the Supermassive Black Hole*, Astrophysical Journal Letters, vol. 875, 1, 2019

Event Horizon Telescope Collaboration, *First Sagittarius A* Event Horizon Telescope Results. VI. Testing the Black Hole Metric*, Astrophysical Journal Letters, vol. 930, 2, 2022

Falcke, H., *The road toward imaging a black hole: A personal perspective*, Natural Sciences, vol. 2, 4, 2022

Falcke, H., Melia, F., Agol, E., *Viewing the Shadow of the Black Hole at the Galactic Center*, Astrophysical Journal, vol. 528, 1, 2000

Fan, X., et al., *A Survey of z > 5.7 Quasars in the Sloan Digital Sky Survey. III. Discovery of Five Additional Quasars**, Astronomical Journal, vol. 128, 4, 2004

Finkelstein, D., *Past-Future Asymmetry of the Gravitational Field of a Point Particle*, Phys. Rev., vol. 110, 965, 1958

Fransson, C., et al., *Emission lines due to ionizing radiation from a compact object in the remnant of Supernova 1987A*, Science, vol. 383, 6685, 2024

Frolov, V.P., Markov, M.A., Mukhanov, V.F., *Through a black hole into a new universe?*, Phys. Lett. B., vol. 216, 1989

Fukue, J., Yokoyama, T., *Colour photographs of an accretion disk around a black hole*, Publ. Astron. Soc. Japan, vol. 40, 1, 1988

Garofolo, D., *Advanced Life Peaked Billions of Years Ago According to Black Holes*, Galaxies, vol. 11, 3, 2023

Gehrels, N., et al., *Ozone Depletion from Nearby Supernovae*, Astrophysical Journal, vol. 585, 2, 2003

Geikie, Sir A., Review of *Memoir of John Michell*, Geological Magazine, vol. 5, 11, 1918

Ghez, A., et al., *High Proper-Motion Stars in the Vicinity of Sagittarius A*: Evidence for a Supermassive Black Hole at the Center of Our Galaxy*, Astrophysical Journal, vol. 509, 2, 1998

Ghez, A., et al., *The First Measurement of Spectral Lines in a Short-Period Star Bound to the Galaxy's Central Black Hole: A Paradox of Youth*, Astrophysical Journal, vol. 586, 2, 2003

Giddings, S.B., Thomas, S., *High energy colliders as black hole factories: The end of short distance physics*, Phys. Rev. D, vol. 65, 5, art. 056010, 2002

Gloudemans, A.J., et al., *Monster Radio Jet (>66 kpc) Observed in Quasar at z ~ 5*, Astrophysical Journal Letters, vol. 980, 1, 2025

Godfrey, B.B., *Mach's Principle, the Kerr Metric, and Black-Hole Physics*, Phys. Rev. D, vol. 1, 1970

Gold, T., *Rotating Neutron Stars and the Nature of Pulsars*, Nature, vol. 221, 1969

Gravity collaboration, *Detection of the gravitational redshift in the orbit of the star S2 near the Galactic centre massive black hole*, A&A, vol. 615, L15, 2018

Haas, R., *Clock comparison using black holes*, Nature Physics, vol. 17, 2021

Hamilton, A.J.S., *The Black Hole Particle Accelerator as a Machine to make Baby Universes*, arXiv:1305.4524, 2013

Hardin, C.L., *The scientific work of John Michell*, Annals of Science, vol. 22, 1966

Hawking, S., *Gravitationally collapsed objects of very low mass*, Monthly Notices of the Royal Astronomical Society, vol. 152, 75, 1971

Hawking, S., *Black Hole Explosions?*, Nature, vol. 248, 1974

Hayes, J.M., et al., *Glimmers in the Cosmic Dawn: A Census of the Youngest Supermassive Black Holes by Photometric Variability*, Astrophysical Journal Letters, vol. 971, 1, 2024

Hayes, L.A., Gallagher, P.T., *A Significant Sudden Ionospheric Disturbance associated with Gamma-Ray Burst GRB 221009A*, Research Notes of the AAS, vol. 6, 10, 2022

Herdeiro, C.A.R., Lemos, J.P.S., *The black hole fifty years after: Genesis of the name*, arXiv:1811.06587v2, 2019

Hughes, D.W., Cartwright, S., *John Michell, the Pleiades, and Odds of 496,000 to 1*, Journal of Astronomical History and Heritage, vol. 10, 2, 2007

Hutton, E., *The Reverend John Michell: A Letter from His Great-grandson*, Antiquarian Astronomer, Issue 3, 2006

Israel, W., 'Dark Stars: the evolution of an idea' in *Three Hundred Years of Gravitation* (ed. Hawking, S., Israel, W.), Cambridge University Press, 1989

Israel, W., *Imploding Stars, Shifting Continents, and the Inconstancy of Matter*, Foundations of Physics, vol. 26, 1996

Jackson, A.A., Ryan Jun, M.P., *Was the Tungus Event due to a Black Hole?*, Nature, vol. 245, 1973

Jia, S., et al., *The Galactic Center: Improved Relative Astrometry for Velocities, Accelerations, and Orbits near the Supermassive Black Hole*, Astrophysical Journal, vol. 873, 1, 2019

Kendall, I.S., et al., *Impacts of UV Radiation from an AGN on Planetary Atmospheres and Consequences for Galactic Habitability*, Astrophysical Journal, vol. 980, 221, 2025

Kerr, R.P., *Discovering the Kerr and Kerr–Schild metrics*, arXiv:0706.1109, 2008

Kovács, O.E., et al., *A candidate supermassive black hole in a gravitationally lensed galaxy at Z ~ 10*, Astrophysical Journal Letters, vol. 965, 2, 2024

Laplace, P.-S., *Beweis des Satzes, dass die anziehende Kraft bey einem Weltkörper so gross seyn könne, dass das Licht davon nicht ausströmen kann*, Allgemeine Geographische Ephemeriden, 1799

Lineweaver, C.H., Patel, V.M., *All Objects and some Questions*, Am. J. Phys., vol. 91, 2023

Lodato, G., Natarajan, P., *Supermassive black hole formation during the assembly of pre-galactic discs*, Monthly Notices of the Royal Astronomical Society, vol. 371, 4, 2006

Lu, R.-S., et al., *Imaging the Supermassive Black Hole Shadow and Jet Base of M87 with the Event Horizon Telescope*, Astrophysical Journal, vol. 788, 2, 2014

Luminet, J.-P., *Image of a spherical black hole with thin accretion disk*, Astron. Astrophys., vol. 75, 228, 1979

Luminet, J.-P., *The Warped Science of Interstellar*, arXiv:1503.08305, 2015

Lynden-Bell, D., Rees, M.J., *On quasars, dust and the galactic centre*, Monthly Notices of the Royal Astronomical Society, vol. 152, 1971

SOURCES

341

McCormmach, R., *John Michell and Henry Cavendish: Weighing the Stars*, British Journal for the History of Science, vol. 4, 2, 1968

Maiolino, R., et al., *A small and vigorous black hole in the early Universe*, Nature, vol. 627, 2024

Maldacena, J., Susskind, L., *Cool horizons for entangled black holes*, Fortsch. Phys., vol. 61, 2013

Marichalar, P., *'This Mountain is It': How Hawai'i's Mauna Kea was 'Discovered' for Astronomy (1959–79)*, Journal of Pacific History, vol. 56, 2, 2021

Matzner, R., Rothman, T., Unruh, B., 'Grand Illusions: Further Conversations on the Edge of Spacetime' in Rothman, T., *Frontiers of Modern Physics*, Dover Publications, 1985

Medeiros, L., et al., *The Image of the M87 Black Hole Reconstructed with PRIMO*, Astrophysical Journal Letters, vol. 947, 1, 2023

Melott, A.L., et al., *Did a gamma-ray burst initiate the late Ordovician mass extinction?*, Int. J. Astrobiol., vol. 3, 2004

Michell, J., *An inquiry into the probable parallax, and magnitude of the fixed stars*, Philosophical Transactions, vol. 57, 1767

Michell, J., *On the means of discovering the distance, magnitude, &c. of the fixed stars*, Phil. Trans. R. Soc., vol. 74, 1784

Minkowski, H., *Raum und Zeit*, Jahresberichte der Deutschen Mathematiker-Vereinigung, 1909

Montgomery, C., Orchiston, W., Whittingham, I., *Michell, Laplace and the Origin of the Black Hole Concept*, Journal of Astronomical History and Heritage, vol. 12, 2009

Morris, M.S., Thorne, K.S., *Wormholes in spacetime and their use for interstellar travel: A tool for teaching general relativity*, American Journal of Physics, vol. 56, 5, 1988

Natarajan, P., et al., *Unveiling the First Black Holes With JWST: Multi-wavelength Spectral Predictions*, Astrophysical Journal, vol. 838, 2, 2017

Natarajan, P., et al., *First Detection of an Over-Massive Black Hole Galaxy UHZ1: Evidence for Heavy Black Hole Seed Formation from Direct Collapse*, Astrophysical Journal Letters, vol. 960, 1, 2024

Nethington, A., *Achieving Philosophical Perfection: Omar Khayyam's Successful Replacement of Euclid's Parallel Postulate*, Lucerna, vol. 14, 2020

Oei, M.S.S.L., et al., *Black hole jets on the scale of the cosmic web*, Nature, vol. 633, 2024

Oort, J.H., Rougoor, G.W., *The position of the galactic centre*, Monthly Notices of the Royal Astronomical Society, vol. 121, 1960

Opatrný, T., Richterek, L., Bakala, P., *Life under a black sun*, American Journal of Physics, vol. 85, 1, 2017

Oppenheimer, J.R., Volkoff, G.M., *On Massive Neutron Cores*, Physical Review, vol. 55, 4, 1939

O'Raifeartaigh, C., Mitton, S., *Interrogating the legend of Einstein's 'biggest blunder'*, arXiv:1804.06768, 2018

Ortega-Rodríguez, M., et al., *The Early Scientific Contributions of J. Robert Oppenheimer: Why Did the Scientific Community Miss the Black Hole Opportunity?*, Phys. Perspect., vol. 19, 2017

O'Sullivan, J., *How we made the wireless network*, Nature Electronics, vol. 1, 2018

Pathria, R.K., *The universe as a black hole*, Nature, vol. 240, 1972

Planck Collaboration, *Planck 2018 results VI. Cosmological parameters*, A&A, vol. 641, 2020

Poplawski, N., *The universe as a closed anisotropic universe born in a black hole*, Gen. Relativ. Gravit., vol. 53, 18, 2021

Raap, A.I., 'Karl Schwarzschild: Leben, Familie und Forschung – eine Übersicht' in *Karl Schwarzschild (1873–1916)*, Universitätsverlag Göttingen, 2017

Raap, A.I., 'Hertzsprung und Schwarzschild in Potsdam: 1910–1916' in *Astrophysik seit 1900*, Nuncius Hamburgensis band 59, tredition, 2024

342   FACING INFINITY

Raymond, A.W., et al., *First Very Long Baseline Interferometry Detections at 870 μm*, Astronomical Journal, vol. 168, 130, 2024

Rees, M., *A better way of searching for black-hole explosions?*, Nature, vol. 266, 1977

Richards, J.L., 'The Geometrical Tradition: Mathematics, Space, and Reason in the Nineteenth Century' in *Cambridge History of Sciences: Volume 5*, Cambridge University Press, 2002

Rindler, W., *Visual Horizons in World Models*, Monthly Notices of the Royal Astronomical Society, vol. 116, 6, 1956

Salazar, J.A., *Multicultural settler colonialism and indigenous struggle in Hawai'i: the politics of astronomy on Mauna a Wākea*, PhD thesis, University of Hawai'i at Mānoa, 2014

Schemmel, M., 'An Astronomical Road to General Relativity' in *In the Shadow of the Relativity Revolution* (ed. Renn, J., Schemmel, M., Wazeck, M.), Max-Planck Institut für Wissenschaftsgeschichte, 2004

Scherrer, R.B., *Gravitational Effects of a Small Primordial Black Hole Passing Through the Human Body*, arXiv: 2502.09734, 2025

Schiavi, R., et al., *Future merger of the Milky Way with the Andromeda galaxy and the fate of their supermassive black holes*, A&A, vol. 642, A30, 2020

Schmidt, M., *3C 273: A Star-Like Object with Large Red-Shift*, Nature, vol. 197, 1963

Schnittmann, J.D., *Life on Miller's Planet: The Habitable Zone Around Supermassive Black Holes*, arXiv:1910.00940, 2019

Scholtz, J., Unwin, J., *What If Planet 9 Is a Primordial Black Hole?*, Phys. Rev. Lett. vol. 125, 2020

Schwarzschild, K., 'Über das Gravitationsfeld eines Massenpunktes nach der Einsteinschen Theorie' in *Sitzungsberichte der Königlich Preußischen Akademie der Wissenschaften*, Verlag der Königlichen Akademie der Wissenschaften, 1916a

Schwarzschild, K., 'Über das Gravitationsfeld einer Kugel aus inkompressibler Flüssigkeit nach der Einsteinschen Theorie' in *Sitzungsberichte der Königlich Preußischen Akademie der Wissenschaften*, Verlag der Königlichen Akademie der Wissenschaften, 1916b

Schwarzschild, K., 'Über den Einfluß von Wind und Luftdichte auf die Flugbahn der Geschosse' in *Sitzungsberichte der Königlich Preußischen Akademie der Wissenschaften*, Verlag der Königlichen Akademie der Wissenschaften, 1920

Shahvandi, M.K., et al., *The increasingly dominant role of climate change on length of day variations*, PNAS, vol. 121, 30, 2024

Shamir, L., *The distribution of galaxy rotation in JWST Advanced Deep Extragalactic Survey*, Monthly Notices of the Royal Astronomical Society, vol. 538, 1, 2025

Sheehan, M., 'Outer Space and Indigenous Security: Sweden's Esrange Launch Site and the Human Security of the Sami' in *Human and Societal Security in the Circumpolar Arctic* (ed. Hossain, K., Martín, J.M.R., Petrétei, A.), Brill Nijhoff, 2018

Sicilia, A., et al., *The Black Hole Mass Function Across Cosmic Times. I. Stellar Black Holes and Light Seed Distribution*, Astrophysical Journal Letters, vol. 924, 2, 2022

Simionescu, A., et al., *Metal-rich multi-phase gas in M 87*, A&A 482, 2008

Simionescu, A., et al., *A Uniform Contribution of Core-collapse and Type Ia Supernovae to the Chemical Enrichment Pattern in the Outskirts of the Virgo Cluster*, Astrophysical Journal Letters, vol. 811, 2, 2015

Siraj, A., Loeb, A., *Searching for Black Holes in the Outer Solar System with LSST*, Astrophysical Journal Letters, vol. 898, 1, 2020

Skulberg, E., Sparre, M., *A Black Hole in Ink: Jean-Pierre Luminet and 'Realistic' Black Hole Imaging*, Historical Studies in the Natural Sciences, vol. 53, 4, 2023

Smolin, L., *Did the Universe evolve?*, Class. Quantum Grav., vol. 9, 1, 1992

Smolin, L., *The fate of black hole singularities and the parameters of the standard models of particle physics and cosmology*, arXiv:gr-qc/9404011, 1994

SOURCES

Sörlin, S., Wormbs, N., 'Rockets and Reindeer: A Swedish Development Pair in a Northern Welfare Hinterland' in *Science for Welfare and Warfare: Technology and State Initiative in Cold War Sweden* (ed. Lundin P., Gribbe J., Stenlås, N.), Sagamore Beach, 2010

Stachel, J., '"What Song the Syrens Sang": How Did Einstein Discover Special Relativity?' in Stachel, J., *Einstein from 'B' to 'Z'*, Birkhäuser, 2002

Stephenson, F.R., Green, D.A., *Was the supernova of AD1054 reported in European history?*, Journal of Astronomical History and Heritage, vol. 6, 1, 2003

Stone, P., *Hutton reconstructed and illustrated, fossils lost and found, dying for science, and a musical survivor*, Magazine of the Edinburgh Geological Society, 52, 2012

Stuckey, W.M., *The observable universe inside a black hole*, Am. J. Phys., vol. 62, 788, 1994

Subrayan, B.M., et al., *Scary Barbie: An Extremely Energetic, Long-Duration Tidal Disruption Event Candidate Without a Detected Host Galaxy at z = 0.995*, Astrophysical Journal Letters, vol. 948, 2, 2023

Suh, H., et al., *A super-Eddington-accreting black hole ~1.5 Gyr after the Big Bang observed with JWST*, Nature Astronomy, vol. 9, 2025

Thomas, B.C., Melott, A.L., *Gamma-ray bursts and terrestrial planetary atmospheres*, New Journal of Physics, vol. 8, 2006

Thorne, K., *Disk-Accretion onto a Black Hole. II. Evolution of the Hole*, Astrophysical Journal, vol. 191, 1974

Tolman, R.C., *Static Solutions of Einstein's Field Equations for Spheres of Fluid*, Physical Review, vol. 55, 4, 1939

Übler, H., et al., *GA-NIFS: JWST discovers an offset AGN 740 million years after the big bang*, Monthly Notices of the Royal Astronomical Society, vol. 531, 1, 2024

Viergutz, S.U., *Image generation in Kerr geometry*, A&A, vol. 272, 1993

Volonteri, M., *Formation of Supermassive Black Holes*, Astronomy and Astrophysics Review, vol. 18, 2010

von Soldner, J.G., *Etwas über die relative Bewegung der Fixsterne*, Astronomisches Jahrbuch für das Jahr 1803 (C.F.E. Späthen, Berlin, 1800)

Wada, K., et al., *Planet Formation around Supermassive Black Holes in the Active Galactic Nuclei*, Astrophysical Journal, vol. 886, 2, 2019

Wada, K., et al., *Formation of 'Blanets' from Dust Grains around the Supermassive Black Holes in Galaxies*, Astrophysical Journal, vol. 909, 1, 2021

Wheeler, J.A., *Our Universe: The Known and the Unknown*, American Scholar, vol. 37, 2, 1968

Wollman, E.R., et al., *Spectral and spatial resolution of the 12.8 micron Ne II emission from the Galactic Center*, Astrophysical Journal, vol. 205, 1976

Wondrak, M.F., van Suijlekom, W.D., Falcke, H., *Gravitational Pair Production and Black Hole Evaporation*, Phys. Rev. Lett., vol. 130, 221502, 2023

Zeldovich, Y.B., Novikov, I.D., *The Hypothesis of Cores Retarded during Expansion and the Hot Cosmological Model*, Soviet Astronomy, vol. 10, 1967

Zhao, X., et al., *Re-estimating the Spin Parameter of the Black Hole in Cygnus X-1*, Astrophysical Journal, vol. 908, 2, 2021

ILLUSTRATION CREDITS

1 Karl Schwarzschild's First World War ID card: Niedersächsische Staats- und Universitätsbibliothek Göttingen
2 Letter from Karl Schwarzschild to Albert Einstein: Niedersächsische Staats- und Universitätsbibliothek Göttingen
3 The structure of a black hole as described by Schwarzschild's formula: Johan Jarnestad/The Royal Swedish Academy of Sciences
4 The gravitational wave detector LIGO in the Hanford desert: LIGO/Caltech/MIT
5 Jean-Pierre Luminet's 1978 scientific illustration of a hot gas swirling around a black hole: Jean-Pierre Luminet
6 Black holes M87* and Sgr A*: Event Horizon Telescope Collaboration
7 Hercules A galaxy and jets: NASA, ESA, S. Baum and C. O'Dea (RIT), R. Perley and W. Cotton (NRAO/AUI/NSF), and the Hubble Heritage Team (STScI/AURA)

INDEX

Page numbers in *italics* indicate illustrations

2012 (film), 233
3C 273 (quasar), 174–5 235

A Hua He Inoa, 216
Abell 2744 (Pandora's Cluster),
  190–91
Africa Millimetre Telescope (AMT),
  167
Agol, Eric, 146
Aldrin, Buzz, 226–7
d'Alembert, Jean le Rond, 15
Allen, Steven, 246
American War of Independence, 14
Andromeda galaxy, 179–80, 246,
  300, 303
Aquila, 101, 102
arcminutes, 156
Arkani-Hamed, Nima, 282
Armstrong, Neil, 28, 226–7
ASTRON (Netherlands Institute for
  Radio Astronomy), 197, 223
Astronomia Nova (Kepler), 39
Astrophysical Observatory Potsdam,
  Berlin, 47
atomic bomb, 108, 109
atomic clocks, 78, 152, 228, 237

Baade, Walter, 107
Babylon, 70
Barish, Barry C., 127, 135
Bekenstein, Jacob, 274–7
Bell Burnell, Jocelyn, 111–12
Bergen, Bishop of, 38
Betelgeuse, 32, 254

Bethe, Hans, 245
BICEP2 telescope, 154
Big Bang, 171–2, 182, 189, 191, 264, 306
black dwarfs, 105
'Black Hole Explosions?' (Hawking),
  268–9
Black Hole Initiative, 150–52, 155, 185
'black hole of Calcutta', 212–13
black holes, *162*, 233–4, 262–3, 310–11
  black hole complementarity, 280
  causality of, 91–3
  collision of, 125–7, 130–31, 137–8,
    180
  creating particles, 265–8
  Earth and, 229–30, 284–5
  electric charge, 117
  entropy of, 276–9
  explosion of, 267–70, 273–4
  formation of, 84–9, 107, 109–11,
    137, 172–3, 191–3, 244, 245, 252
  image of, 141–4, 146–62, 166–8
  International Celestial Reference
    Frame (ICRF), 228
  magnetic fields, 147, 154, 165–6
  mathematical formula for, 46,
    48–9, 81, 83–4, 230
  navigation and, 238–42
  new universes and, 296–9, 300
  origin of name, 212–13
  origins of life and, 243, 249–51,
    254–61, 298, 311–12
  perceived experience of, 1–6, 94–5
  primordial black holes, 177, 178–9,
    183, 284–5

346 FACING INFINITY

black holes (*cont.*)
relativity of time and, 90–91
rotation of, 112–19, 120, 130, 147,
165, 175, 265–7, 296–7
shadow of, 146–7, 155, 156, 158–9,
166
size and mass of, 89–90, 94, 117,
163–4, 186–8, 299
structure of, 94
supermassive black holes, 138,
148, 162, 169–76, 178–9, 180–81,
187–8, 189, 245–6, 248–9, 255–6,
261, 308–9
universe as a black hole, 299–307
wormholes and, 293–4
see also event horizon; galactic
jets; gravitational waves; inner
horizon; quasars; singularity
as well as individual names of
black holes
blanets, 257–60
Board of Longitudes, 203, 229
BOAT ('Brightest of All Time'),
251–2
Bogdán, Ákos, 191
Boltzmann, Ludwig, 275–6, 277–8
Bolyai, Farkas, 71–3
Bolyai, János, 71–4
Bower, Geoff, 25, 43, 144, 215, 217
Brahe, Tycho, 35–9, 42
Bras, Albert Le, 208–9
Bremer, Michael, 294–5
Brezhnev, Leonid, 264
Brief History of Time, A (Hawking),
271
Broderick, Avery, 311
Brown, Robert, 29
Bureau des Longitudes, 174

Canadian Space Agency, 188
Canterbury College, Christchurch,
113
Capella, 227

Carroll, Lewis (Charles Lutwidge
Dodgson), 75
Carter, Brandon, 143
Case, Pua, 221
Cavendish, Henry, 21, 230
Center for Astrophysics | Harvard &
Smithsonian, 148, 191
Cepheid variable, 301
CERN, 281, 282
Chadwick, James, 107
Chalmers University of Technology,
225
Chandra X-ray Observatory, 191, 192
Chandrasekhar, Subrahmanyan, 105,
106, 206
'Le Christ aux oliviers' (Nerval), 143
climate change, 224–5, 233, 237,
240–41
colonialism, 197–215
'Concise Outline of the Foundations
of Geometry, A' (Lobachevsky),
73
continental drift, 228–34
Cook, James, 202–4
Copernicus, 14, 36, 39, 303
corps obscurs, 22, 24
cosmic background light, 303
cosmic censorship hypothesis
(Penrose), 117
cosmic event horizon, 305–7
Cosmic Explorer (USA), 138
cosmic gas clouds, 173, 183
cosmic radiation, 283
cosmological natural selection, 298
Cosmos (TV documentary), 243
COVID-19 pandemic, 135
Crab Nebula, 106, 112
CSIRO Division of Radiophysics, 269
Curie, Marie, 28
Curtis, Herbert, 164
curvature of spacetime, 75, 80, 84,
90, 95, 146, 174, 258–9, 296
Cygnus, 119

INDEX

Cygnus X-1, 119–20, 235

Daily Mail, 280–81, 283
dark energy, 305–6, 308
dark matter, 177–9, 304
dark stars, 15–20
Darwin, Charles, 288
De Fries, John, 215
Dempsey, Jessica, 197–202, 215–19, 221–3, 311
Dicke, Robert, 213
Doeleman, Sheperd, 148–9, 150–53, 155, 158, 160, 161, 167, 220, 295
Doppler effect, 142, 155
Dwyer, Sheila, 137–8, 139

Earhart, Amelia, 28
Earth, 14, 32, 40, 74, 227–31, 236–46, 251
rotation of, 236–9
Earth's Shifting Crust (Hapgood), 233
earthquakes, 229–30
Eddington, Arthur, 107, 146, 205–6
Einstein, Albert, 3, 48–9, 50–51, 53–63, 65–70, 74, 75–81, 82, 83–4, 85, 95–7, 102, 107, 114, 118, 120, 137, 146–7, 163, 167, 205–6, 233, 265, 272, 292, 301–2, 310
see also general theory of relativity; quantum mechanical description of light; special theory of relativity
Einstein, Elsa, 272
Einstein, Hermann, 50–1
Einstein, Jakob, 51
Einstein, Pauline, 50
Einstein Telescope (Europe), 138
Einstein–Rosen bridge, 292
El Niño weather system, 236
electric lights, 51, 61
Elements, The (Euclid), 70
Enceladus, 257
Englert, François, 283

Enlightenment, 14, 15
entropy, 275–8, 279
Epstein, Jeffrey, 287
ergosphere, 115–16
escape velocity, 18–19
Escher, M.C., 75
Esrange, Sweden, 208
Euclid, 70–72
Euclid's fifth postulate, 71–2
Europa, 257
European Southern Observatory (ESO), 101
European Space Agency (ESA), 188, 288
European Space Research Organisation (ESRO), 208–9
event horizon, 3, 85–7, 92, 93, 95, 97, 116–17, 161, 292, 307–8
Event Horizon Telescope (EHT), 25, 144, 149–55, 156–9, 160, 161, 166, 168, 207, 215, 220, 228, 256, 293–5
expansion of space, 300–303, 306–7, 309
Extremely Large Telescope (ELT), Chile, 193

Faber, Sandra, 201
Falcke, Heino, 145–7, 149, 150, 157, 158, 159, 161, 164, 167
Faust (Goethe), 103–4
Finkelstein, David, 92
first law of thermodynamics, 275
First World War (1914–18), 46, 48–9, 76–7
Flamm, Ludwig, 291–2
Fourier analysis, 153
France, 14
Franklin, Benjamin, 21, 22
French Revolution (1789–99), 14
Frolov, Valeri, 295–7
fusion, 102–5

Gaia, space observatory, 101, 102, 238
Gaia BH1, 102, 288
Gaia BH3, 101–2, 120
galaxies, 169–70, 187
  galactic collision, 179–80
  galactic jets, 164–6, *165*, 169, 248, 255–6
  habitable zones, 254–7
Galileo Galilei, 33, 66, 211
Gamma Cassiopeia, 227
gamma-ray bursts, 251–3, 254
Gamow, George, 302
Gauss, Carl Friedrich, 73–5
general theory of relativity (Einstein), 65, 77–9, 82, 83, 96, 137, 146–7, 167, 265–6, 273–4, 276–7, 279, 293
Genzel, Reinhard, 24, 31, 42–3, 44, 89, 163
geocentric model, 36
geometry, 68–75
  non-Euclidean, 69–76
'geopoetic essay, A' (Hess), 234
Germany, 37
Ghez, Andrea, 24, 27–8, 29–31, 33–5, 42–3, 44, 89, 145, 163, 164, 199
Global Millimeter Very Long Baseline Array, 166
Global Positioning System (GPS), 77–8, 238, 241–2
Glossopteris fern, 231–2
GN-z11 galaxy, 189
Goethe, Johann Wolfgang von, 103
gravitational waves, 121–7, *123*, 129–31, 132–40, 180–82
gravity, 16–19, 45, 48, 64, 65–8, 77, 85
Greenland Telescope, 207
Greenwich Mean Time, 203
Grossmann, Marcel, 68–70, 76
Guiana Space Centre, Kourou, 188, 209–10

Haas, Rüdiger, 226, 227, 236–41
Halley's comet, 186
Hanford desert, 121, 139
Hapgood, Charles, 233
Harmonices Mundi (Kepler), 39
Harrison, B. Kent, 110
Hartmannswillerkopf, Vosges mountains, 290–91, 310
Harvard University, 151
Hawai'i, 197–205, 215–7, 219–23, 281
Hawking, Jane *see* Wilde, Jane
Hawking, Stephen, 95, 178, 186, 262–8, 270–73, 283, 284–5, 286–9, 312
Hawking radiation, 269–70, 272–4, 276–7, 280, 282, 308, 309
Haystack Observatory, 228–9, 235
heliocentric model, 36
Hercules A galaxy, *165*
Hess, Harry, 234
Hewish, Antony, 111
Higgs, Peter, 283, 287
Higgs Boson, 281, 283
Hilbert, David, 76
holographic principle, 279
Homer, 226
Hooft, Gerard 't, 279
Hubble, Edwin, 300–302, 305
Hubble Space Telescope, 186, 189, 201
Hulse, Russell A., 134
Hume, David, 54–5
Hutton, Scot James, 14
hydrogen bomb, 109

Ibn Butlan, 106
'Imiloa Astronomy Center, 216
Independent, 299
infinity, 95
information paradox, 273–4, 280
inner horizon, 116
Institut d'Astrophysique de Paris, 170

INDEX

intergalactic medium, 246–7, 249
Intergovernmental Panel on Climate
  Change (IPCC), 224, 240
International Astronomical Union
  (IAU), 219
Interstellar (film), 90–1), 292–3
IRAM 30-meter telescope, 150
Israel, Werner, 233
Issaoun, Sara, 144–6, 149–60, 162–4,
  166–8

J0529-4351 (quasar), 175
James Clerk Maxwell Telescope
  (JCMT), Maunakea, 197–8,
  215
James Webb Space Telescope
  (JWST), 188–92, 209, 252,
  256–7, 299, 307
Juno, 182
Jupiter, 40, 182

K2-18b (exoplanet), 257
KAGRA (Japanese gravitational
  wave detector), 135
Kalaniʻōpuʻu, King of Hawaiʻi, 203
Kamehameha I, King of Hawaiʻi,
  204
Kattegat strait, 225
Kavli Institute for Particle
  Astrophysics and Cosmology,
  California, 246
Kazan Messenger, 73
Keck, William Myron, 30
Keck telescopes, 33, 43, 199
Keliiheleua, Sam, 222
Kepler, Johannes, 36–42, 48
  Kepler's First Law, 41
  Kepler's formula, 104
  Kepler's Second Law, 41
  Kepler's Third Law, 40–41, 42
Kerr, Roy, 112–16, 118–19, 120, 161, 174,
  175, 211, 296–7

Khayyam, Omar, 71
Kilauea, Hawaiʻi, 25
kilonova, 137
Kimura, Larry, 215–16
Kimura, Leslie Kaʻiu, 217–21
Kristian IV, King of Denmark, 36
Kuiper, Gerard, 198

Lambda Aquilae, 102
land rises, 239–40
Landry, Michael, 124–7, 135–6, 139
Langer Max, 230
Laplace, Pierre-Simon, 14–15, 22–4,
  45, 85, 174
Large Hadron Collider (LHC), 281–4
Larrue, André, 290
Laser Interferometer Space Antenna
  (LISA), 138, 180, 193, 260
law of gravitation (Newton), 65
Le Verrier, Urbain, 80–81
Leavitt, Henrietta Swan, 301
Lemaître, Georges, 302
light, 50, 51–2, 60
LIGO (Laser Interferometer
  Gravitational-Wave
  Observatory), 122–7, *123*, 131,
  134–40, 180, 181
Liliʻuokalani, Queen of Hawaiʻi,
  204–5
little red dots, 189–90
Lobachevsky, Nikolai, 73–5
Local Group, 246
Lodato, Giuseppe, 187
LOFAR, 248
Lorentz, Hendrik, 60
Lovecraft, H.P., 75
Luminet, Jean-Pierre, 141–4, *142*, 147,
  153, 155, 161, 295
Lundmark, Knut, 106, 300, 302
Lynden-Bell, Donald, 28

M87 galaxy, 143, 147, 164, 246, 260
M87* black hole, 147, 149, 151, 154–9,
    160–3, 162, 164, 166–7, 168, 169,
    215–20, 223, 242, 260, 313
Machiavelli, Niccolò, 134
Magorrian, John, 186
Maldacena, Juan, 279
Manhattan Project, 109
MANIAC (Mathematical Analyzer
    Numerical Integrator and
    Automatic Computer), 109–10
Marić, Mileva, 55, 272
Markov, Moisey, 295–7
Mars, 40
Mason, Elaine, 272
Maunakea, Hawai'i, 25–6, 30, 33–4,
    43, 197–202, 205, 215, 221–3, 311
Maunaloa, Hawai'i, 27
Max Planck Institute of
    Gravitational Physics,
    Germany, 281–2
Maxwell, James Clark, 51–3
McGill University, Montreal, 145
Melia, Fulvio, 146
Melott, Adrian, 252–3
Mercury, 40, 80
Messier, Charles, 147
Meudon Observatory, 141
Michell, John, 12–24, 29, 44, 45, 85,
    101, 174, 203, 229, 310
Mid-Atlantic Ridge, 234
military involvement in astronomy,
    210–12
Milkomeda, 179, 180
Milky Way, 11, 14, 28–9, 32–3, 34,
    43–4, 101, 170, 179–81, 188, 238,
    301, 303–4
Milner, Yuri, 287
Minkowski, Hermann, 62–3
Monoceros, 288
Moon, 32, 236–7
Mount Graham, Arizona, 149

Mukhanov, Viatcheslav, 295–7
Museum of Modern Art, New York,
    159

Namur, Belgium, 45
NASA, 27, 102, 188
Natarajan, Priyamvada, 184–91, 192
National Institute of Standards and
    Technology (NIST), 77
National Science Federation, 134
Nature journal, 284
Neffe, Jürgen, 61
Neil Gehrels Swift Observatory, 251
Nerval, Gérard de 143
neutron stars, 106–7, 110, 111–12,
    130–31, 134, 136–7, 244, 258
New York Times, 282
Newton, Isaac, 17, 42, 60, 64–6,
    80–81, 85, 288
    Newton's theory of gravity, 174
Next Generation Event Horizon
    Telescope (ngEHT), 144, 167
NGC 4993 galaxy, 136
'NIST Clock Experiment
    Demonstrates That Your Head
    Is Older Than Your Feet', 77
Nixon, Richard, 264
Nobel Prize, 65, 89, 105, 111, 118, 127,
    134, 154, 245, 283
non-Euclidean geometry, 69–76
Northern Extended Millimeter
    Array (NOEMA), 294
nuclear weapons, 214

O'Sullivan, John, 269–70
Odyssey (Homer), 226
Oei, Martijn, 248–9
'On the Means of Discovering the
    Distance, Magnitude, &c. of
    the Fixed Star' (Michell), 21
Onsala Space Observatory, Sweden,
    225, 228–9, 235, 241

INDEX

Opatrný, Tomáš, 258
Oppenheimer, Robert, 107–9, 211
Ordovician–Silurian mass
    extinction, 252–3
Origin of Continents and Oceans, The
    (Wegener), 232
Orion, 32, 254
Ortelius, Abraham, 231

Pangaea, 232
Panthalassa, 232
Papapetrou, Achilles, 118
parallax measurement, 15–16
Paris Observatory, 170, 173–4
Parque de las Ciencias, Spain, 144
Pascal, Blaise, 2
Pauli exclusion principle, 104
Penrose, Roger, 24, 117, 118, 264
Perseus cluster, 247
*Philosophiæ Naturalis Principia
    Mathematica* (Newton), 17
Phoenix A*, 176
Pioneer 11, 102
Planet Nine, 285
Poincaré, Henri, 60
Polytechnic Institute, Zurich, 53, 54, 55
Porphyrion jet, 248–9
Pōwehi *see* M87*
Prague, Czech Republic, 35
Príncipe, West Africa, 205
Proxima Centauri, 31–2, 181
Prussian Academy of Sciences,
    Berlin, 48, 77, 81
Pulsar Timing Array, 181–2
pulsars, 111–12, 181–2

quantum field theory, 265–6
quantum gravity theories, 96
quantum mechanical description of
    light (Einstein), 59
quantum mechanics, 80, 96, 97,
    104–5, 265–6, 273, 279, 293

quasars, 174–6, 226, 227–9, 235, 236,
    238–9, 241–2
    system of cosmic coordinates, 236
    see also black holes: International
    Celestial Reference Frame
    (ICRF)

Radboud University Nijmegen, 145
La Recherche, 143
Recorde, Robert, 71
redshift, 87, 302
Rees, Martin, 28, 186, 269
Reitze, David, 127
Riemann, Bernhard, 74
Rigel, 227
Rijksmuseum, Amsterdam, 159
Rindler, Wolfgang, 85
Rolland, Romain, 76
Romania, Conservative Party, 281
Rosen, Nathan, 292
rotating spacetime, 115–17
Royal Society, 13, 21–2
Royal Swedish Academy of Sciences,
    Stockholm, 24
Rudolf II, Holy Roman Emperor, 35

So-2 (star), 34–5
Sagan, Carl, 243
Sagitta, 251
Sagittarius, 34
Sagittarius A*, 11*, 29, 42–3, 44, 89,
    94, 95, 101, 145, 147, 149, 151,
    154–5, 162–4, 166–7, 168, 169,
    179, 183, 188, 199, 256, 294, 313
Saturn, 40
'Scary Barbie' (ZTF20abrbeie), 253
Schild, Alfred, 113–15
Schmidt, Maarten, 174
Schwarzschild, Karl, 45–8, 46, 49, 81,
    82–8, 93–4, 95–7, 118, 163, 186,
    211, 230–31, 232, 235, 290–92, 310
Schwarzschild metric, 49, 87, 90,

93–4, 94, 95, 97, 112, 156, 291–2, 299, 304–5
Schwarzschild radius, 87–9
Schwarzschild, Martin, 186
Science Absolute of Space, The, (Bolyai), 73
Scotland, 14
sea levels, 239–41
second law of thermodynamics, 275, 276
Second World War (1939–45), 97
Seven Years' War (1756–63), 13–14
Sikku, Ol-Johán, 209
Simionescu, Aurora, 243–51, 254–6, 259–61
Simons, Douglas, 217
Sin-bel-apli, 70
singularity, 5, 93–6, 116–17, 306–7
Siraj ud-Daulah, 212
Sirius, 74–5, 104
Sirius B, 104
Skanör-Falsterbo, Sweden, 239
Slipher, Vesto, 302
Smolin, Lee, 297–9
Snyder, Hartland, 107–9
solar eclipse, 146, 205
solar system, 11, 32, 40–41, 52, 182, 238
    anomaly in, 80–81
    composition of, 247–8, 250
    structure of, 36–7
Soldner, Johann Georg von, 24
sound waves, 129–30
Southern Cross, 43
Soviet Union, 264
spacetime, 62–3, 68–9, 75–6, 79–80, 92, 93, 95, 110, 115, 116, 180, 238, 266, 291–2
'spaghettification' (Hawking), 95
special theory of relativity (Einstein), 59–60, 62, 65–6, 79–80, 86, 92
speed, 51–2, 59

speed of light, 52–4, 56, 60, 79
Square Kilometre Array (SKA), 223
SRON Netherlands Institute for Space Research, Leiden, 243
Starobinsky, Alexei, 265–7
stars, 172–3, 226–7, 243–4
    death of, 103–11
    supermassive, 173
    see also black dwarfs; dark stars; neutron stars; pulsars; supernovas; white dwarfs *as well as* names of individual stars
Stjerneborg Observatory, Ven, Sweden, 35–6
subatomic vacuum, 265
Submillimeter Array, 25, 215
Submillimeter Telescope, 150
Sun, 14, 32, 39–40, 74, 102–4, 236
    solar eclipse, 146
supermassive stars, 173
supernovas, 106, 107, 110, 244–5, 254
Susskind, Leonard, 279
Suzaku, X-ray telescope, 246, 247
Swiss Federal Institute of Technology, 69

Taurus, 20, 106
Taylor, Joseph H. Jr, 134
tectonic plates, 234–5, 239–40
Tenerife Event-horizon Antenna, 167
Tharp, Marie, 234
thermodynamics, 274–8
Thirty Meter Telescope (TMT), 198–202, 215, 221–2
Thorne, Kip, 77, 117, 120, 127, 133, 135, 264, 293
Thornhill, Yorkshire, 12, 20–21, 23
    Thornhill Rectory, 13
'Through a Black Hole Into a New Universe?' (Frolov, Markov & Mukhanov), 296

INDEX 353

Thule Air Base, 207
time, 54–5, 57–9, 90–91
Tonantzintla 618, 176
Tong, Iwakelii, 220
Treatise of Human Nature, A
(Hume), 55
Tunguska river, Russia, 284–5

UHZ-1 galaxy, 190–91, 192–3
universe, 17, 64, 75, 127, 133–4, 138,
154, 170–79, 180, 187–90, 220–21,
227, 249, 250, 268, 283
as a black hole, 299–307
end of, 308–9
University of California, Los
Angeles, 28
University of Cambridge, 12
University of Göttingen, Germany,
45
Uraniborg Castle, Ven, Sweden, 35–6

V616 Monocerotis, 288–9
vacuum, 265–6, 305, 307
Vangelis, 288–9
Venus, 40
Very Large Telescope, Chile, 31, 101
very-long-baseline interferometry
(VLBI), 148, 149, 228, 236
Virgo (European gravitational wave
detector), 135–6
Virgo cluster, 147
virtual particles, 265–8
Volonteri, Marta, 170–75, 176,
179–84, 185, 192–3
Vulcan, 81

W.M. Keck Observatory, 30
Wada, Keiichi, 257
Wakano, Masami, 110
War On Mars, The (Kepler), 38
Weber, Joe, 132
Wegener, Alfred, 230–33, 235

Weiss, Rainer, 127, 128–29, 131–3, 135,
139–40
Westford Radio Telescope, 228–9,
235
Wheeler, John Archibald, 109–10,
213–14, 292–3
white dwarfs, 104–5, 107, 206
Wi-Fi, 270
Wilczek, Frank, 287
Wilde, Jane, 263, 264, 271–2, 286
Wilson, John Tuzo, 235
WLAN (Wireless Local Area
Network), 269
Wohl, Liebmann, 87
Wolf, Christian, 175
wormholes, 292–4

X-ray
emissions, 119
Imaging and Spectroscopy
Mission, 260
photons, 191

Zeldovich, Yakov, 211, 264–7
Zimmer, Hans, 91
ZTF20abrbeie, 253
Zuckerberg, Mark, 287
Zwicky, Fritz, 107

ABOUT THE AUTHOR

JONAS ENANDER has a PhD in physics, having conducted research in cosmology and astrophysics. He has worked as a science communicator at the European Southern Observatory in Garching and the Oskar Klein Centre in Stockholm, as well as participated in the construction of the IceCube observatory at the South Pole, Antarctica. He regularly writes about physics and astronomy for various popular science magazines and also hosts two podcasts about science – *Spacetime Fika* in English and *Rumtiden* in Swedish – where he meets scientists to discuss what we know about the universe and how we know it. *Facing Infinity* is his first book.

facinginfinity.com | jonasenander.com